· KEEPING ·
CHICKENS
IN YOUR GARDEN

· KEEPING ·
CHICKENS
IN YOUR GARDEN

A PRACTICAL GUIDE TO RAISING CHICKENS, DUCKS, GEESE AND
TURKEYS IN YOUR BACKYARD, WITH OVER 400 PHOTOGRAPHS

FRED HAMS

southwater

This edition is published by Southwater
an imprint of Anness Publishing Ltd
108 Great Russell Street
London WC1B 3NA
info@anness.com

www.southwaterbooks.com; www.annesspublishing.com

If you like the images in this book and would like to investigate using them for publishing,
promotions or advertising, please visit our website www.practicalpictures.com for more information.

A CIP catalogue record for this book is available from the British Library.

Publisher: Joanna Lorenz
Photographers: Mark Winwood (pp19–109) and Mark Wood (pp90–95)
Illustrator: Stuart Jackson-Carter
Designer: Nigel Partridge
Production Controller: Mai-Ling Collyer

Previously published as part of a larger volume, *The Practical Guide to Keeping Chickens, Ducks, Geese & Turkeys*

PUBLISHER'S NOTE
Although the advice and information in this book are believed to be accurate and true at the time of going to press, neither the authors
nor the publisher can accept any legal responsibility or liability for any errors or omissions that may have been made nor for any
inaccuracies nor for any loss, harm or injury that comes about from following instructions or advice in this book.

Thanks to the following libraries and individuals for permission to use their images:
Alamy: 8t, 9tr, 10t, 36t, 40tr, 56, 65tr, 90tr, 110, 113b, 117tr, 118r, 119b, 121b, 124t. Ardea: 117b. Clare and Terry
Beebe: 33t, 78bc, 78br, 79tl, 79tc, 79tr, 79b, 83tc, 83tr, 94tc, 131bl, 131bc. Bridgeman Art Library: 9tl. British Hen
Welfare Trust: 34–35. Corbis: 9b, 11b, 122b, 125t, 130b, 131t. Fotolia: 57, 115, 128bl. GAP: 117t. iStockphoto: 10b,
21br, 123t, 124b. Dave Scrivener: 33b, 91br, 131r. Roger Sing: 88b. Superstock: 56. Philip Lee Woelf: 76b.

CONTENTS

INTRODUCTION

There are many reasons to keep poultry, and plenty of people take pleasure in looking after these rewarding birds. Like any other pet, poultry will require your time and attention. They will also need space to exercise, a secure, draught-free home, and daily fresh food and water.

A regular source of truly fresh, quality-assured eggs is the motive for an increasing number of people to keep poultry. The appeal may be driven in part by concern about food production and animal welfare. Knowing where our food comes from and what it, in turn, has eaten are important factors in many buying decisions today, as are knowing how it has been treated and whether it was reared according to its natural instincts. By keeping our own hens, we know exactly what they have been fed. We have also taken responsibility for providing them with a suitable environment in which to live and thrive. It is not necessary to keep a rooster if you require eggs to eat. A rooster is only essential if you want fertile eggs, either to hatch and nurture the chicks, or to pass to someone who does.

▼ *If you choose the correct breed, hens can make ideal pets for children.*

▶ *If you keep enough hens, they will produce a surplus of eggs that you can sell or give away.*

Keeping poultry as productive pets

Nurturing any bird or animal, whether it is poultry, a cat or dog, makes us live life at a different pace, partly because the animal is dependent upon us for its welfare and requires our time and attention. Poultry can add immensely to our enjoyment of time spent in the garden as well as promote the garden economy. Their comical and inquisitive behaviour can be highly entertaining, and many will become tame enough to handle, even by young children. Taking time to enjoy the presence of the pet and its behaviour, to care for it and reap the rewards, are the pleasures of owning poultry. Additionally, watching a hen turn broody, then sit and hatch chicks naturally, can provide a family with a magical insight into part of life's cycle.

Keeping pure and rare breeds

For some, owning a pure breed of poultry is nostalgic – the equivalent of owning a piece of living history, and the opportunity to keep an old-fashioned breed that may have been familiar to our grandparents. Usually, eggs are a secondary consideration for this type of poultry keeper, since many of the old pure breeds are not such prolific egg-layers as today's modern hybrid birds.

There will be others who, having seen pictures of exotic-looking fowl in early poultry books, are amazed to find that the descendants of those

fowl are still bred. The popularity of exotic breeds is currently booming. Often purchasing such an unusual bird creates a sense of duty to perpetuate and select future generations of fowl in order to ensure the survival of the breed in as good order as possible.

Keeping poultry for exhibition

Few beginners keep poultry with the sole intention of exhibiting them. However, those who have chosen a pure breed because of its shape, character or feather markings will be inclined to compare birds that they have bred themselves with others of the same breed or variety. Exhibiting can add much to the poultry breeder's experience. Poultry owners who exhibit at their local show meet and exchange notes with others, and will often try to rear a bird as close as possible to the written standard for their breed.

Keeping poultry as a business proposition

Poultry keeping can become a small-scale business, should you decide to keep poultry as breeding stock. The fertile eggs, or the resultant young birds, can be sold on either for hatching, for pets, or for future egg-laying. However, poultry keepers engaged in breeding poultry and artificially incubating them to sell for profit carry a responsibility to future generations of fowl and their keepers.

Some breeders may set out to supply a local requirement for free-range, or organically fed, poultry and their eggs. Stringent regulations govern such activities. Unless you have bred the pullets (young birds) for such a business from your own stock, it is likely that you will need to buy them in. Many such birds may have been bred and reared in intensive bio-secure conditions. Finding a supplier that has bred birds in a way that you find acceptable will take time.

Very dark brown eggs, which are perceived to be healthy, or blue eggs, which seem unusual and therefore an appealing business prospect, are produced almost exclusively by pure breed, genuine utility strains. However, utility strains of pure breeds are rare. Today most pure breeds are reared and selected for exhibition purposes at the expense of their utility value. Pure breeds with utility

▲ *As well as adding interest to the garden, some enjoy exhibiting their poultry at local or national shows.*

▼ *Collecting the eggs is always an appealing job. Everyone can find magic in the discovery of a newly laid egg.*

value do not lay eggs in such large numbers as hybrid birds, and generally cost more to keep; therefore, the eggs produced will necessarily require a higher selling price than standard hybrid eggs. Utility strains of pure breeds are rarer than exhibition strains, adding to the difficulty of locating breeding stock. Few breed clubs have interest in the utility roots of the breeds they represent. However, club utility breeders are likely to be aware of other vendors, and may be able to provide you with details of breeders in your area. Would-be breeders may have to rely on the expertise of just a few utility enthusiasts.

WHAT TO LOOK FOR IN THIS BOOK

This volume is intended as a reference for those who are considering keeping chickens, or who already own poultry and wish to learn more. It is also suitable for those who wish to develop their interest in poultry breeding or exhibiting. The book begins with a historical perspective on the evolution of wild jungle fowl to the domestic chicken breeds that we recognize today. Various breeds are described, as well as their distinctive features such as combs, feather structure, patterns and colours. The life cycle, which has three separate phases, is outlined, and what to expect during development.

First, all the different aspects of setting up as a poultry keeper are explained, with information to help clarify the most likely reasons for buying poultry, whether to choose a pure breed or a hybrid, and the benefits of each. Buying housing is a significant financial outlay, and choosing the correct type for the breed of bird is essential: dedicated chapters explain all the choices available. Responsible ownership is key for healthy poultry, and every aspect of poultry care and management is looked at here in detail, including feed types, health issues, poultry behaviour and egg problems. For keepers who wish to develop their interest further, there is expert advice on setting up a breeding program, incubating eggs, rearing chicks under a broody hen, and caring for young birds. For poultry owners interested in exhibiting their prized breed or indeed, eggs, there is advice on what to expect at the poultry show.

▲ *Hens in an ideal outdoor environment.*

THE EVOLUTION OF DOMESTIC FOWL

Domestic fowl belong to the order Galliformes, which includes guinea fowls, peafowls, pheasants, turkeys, grouse, chickens and quail, among others. These are all heavy, ground-feeding birds. Domestic and wild jungle fowl are descended from the genus *Gallus.*

Four species of wild jungle fowl have been considered as ancestors of all domestic chicken breeds (*Gallus domesticus*): *Gallus gallus* (red jungle fowl), *Gallus lafayette* (Ceylon or Sri Lankan jungle fowl), *Gallus sonneratii* (gray jungle fowl) and *Gallus varius* (Java or green jungle fowl). The wild jungle fowl still exists, although it is known to be endangered in its natural range. It is distinctly recognizable as poultry, but is small and light in weight compared to many modern breeds.

The main issue dividing poultry breeders and scientists is whether or not a single species of wild jungle fowl is the progenitor of all domestic fowl. Charles Darwin was one of many scientists to favour the view that the red jungle fowl is the

▼ *It is thought most likely that poultry originated in Asia and that they were spread out across the globe by traders, who carried them for food and to sell.*

▲ Gallus lafayette, *the Sri Lankan jungle fowl, has been shown to produce fertile offspring when crossed with domestic fowl.*

ancestor of all modern poultry. Poultry fanciers (the name given to all those who keep and breed poultry, often on a small scale), particularly those keeping the older strains of game fowl, are more likely to favour the view that domestic poultry have more than one jungle fowl species as an ancestor.

Early migration of poultry

Domestic breeds of poultry have been known to humans for thousands of years. It was originally assumed that domestication of poultry could be traced to the Indus Valley, a vast area that covered parts of Afghanistan, Pakistan and northwest India, around 2000BC. Cockfighting was known to have taken place there at that time and continued to be a significant reason for keeping poultry up to the mid-19th century. It was thought that domesticated poultry spread to Europe via Persia (modern-day Iran) and Asia Minor from this region. Archaeological evidence has since shown that domestication of poultry occurred in China around 6000BC. The 4,000 additional years would have provided plenty of opportunity for different types of fowl to develop

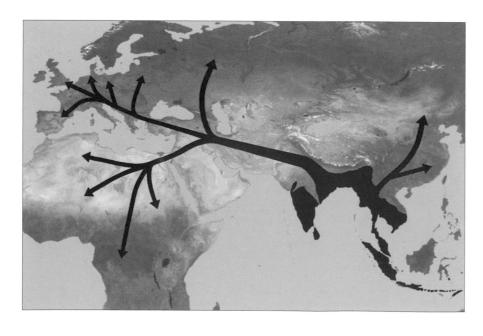

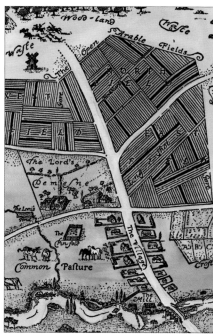

▲ *When the Romans conquered Britain, they would have brought chickens with them. These breeds may have included five-toed fowl that could have been the ancestors of Dorkings.*

within a sophisticated civilization, and with significant differences in size, shape, visual appearance and characteristics.

Poultry spread to America in the 17th century, taken by emigrants from Europe. Turkeys, which are native to that continent were brought back to Europe from America at around the same time.

The development of farming

From the 13th to the 19th centuries most of the developing world lived an agrarian life. Peasants were tied to a manorial system, in which they were protected by the lord of the manor. Strips of land around the estate were parcelled out to workers who lived on

▶ *The traditional farmyard scene in which poultry co-existed with other livestock developed from the feudal system of land tenure.*

the produce they grew. Each worker had several strips of land ensuring that each got a share of the good land as well as that which was less productive. In return for the land, peasants paid a tax either giving a portion of their labour or produce to support the manor house. Additionally common land, not owned by the manor, was used by everyone for growing crops and grazing animals. With the agricultural

▲ *During the Middle Ages game birds were kept for cockfighting. Tenants may have kept animals including poultry on the strips of land that they tended as well as on common land.*

revolution, the system of working strips of land disappeared as fields became enclosed in order to become more productive, and to keep grazing animals within the boundaries. As understanding of crop rotation grew,

the land was made more productive. Labourers were given land around their cottages, rather than strips of land, and from this a system of subsistence farming developed. Each peasant would farm his own land and keep on it the animals and fowl he required to keep his family. With improved production, surplus crops could be sold, changing the nature of the relationship between landowner and tenant.

The 19th century

It was in the 19th century that animal husbandry developed on a greater scale, producing surpluses to feed the growing numbers of people migrating to towns and cities as the industrial revolution took hold. The poultry industry as we know it evolved at this time. Prior to this date, apart from keeping a few birds for eggs, feathers or for the pot, most owners kept birds for cockfighting.

Cockfighting

Keeping birds for sport-fighting was once universally popular the world over. Julius Caesar wrote that the Britons kept fowl for pleasure and as a diversion. This "diversion" is

understood to mean cockfighting. Early references to cockfighting suggest that for most of the last thousand years it has had a dominant influence on how poultry was selected and bred.

The attendance at cockfights may have included all classes of society, but breeding and creating fighting strains remained under the patronage of the nobility as part of the feudal system. Initially these breeding programs would have built on the inherent hardiness of the fowl.

▲ Smallholdings developed at the end of the feudal system, with each tenant farmer tending his own fowl and animals on his own land.

Bred for fighting

Britain had early contact with India, and Indian breeds with more muscular stature were crossed with English breeds to develop offspring with a heavier build. Families of cockfighting game developed in the hands of individual breeders. Over time these differences effectively became separate breeds, as breeders selected game for specific features.

Game fowl bred for sport were reared free range and at worst died in maturity. It was probably between the 16th and early 19th centuries that cockfighting reached its greatest refinement. Fights, brawls and unruly behaviour at cockfights had as much to do with cockfighting being made illegal, as any cruelty considerations. A series of laws passed between 1835 and 1849 finally ended legal cockfighting in Western Europe.

◄ An ill-matched and probably impromptu cockfight – many still take place in Southern and South-East Asia.

Poultry mania

In 1843, a huge poultry breed was brought to Britain from Asia by British naval officer Sir Edward Belcher. Belcher had just returned from a mission to survey the waters around China, but on arrival, he found that the area had already been surveyed. He therefore sailed on into Indonesian waters, making a fleeting visit to Vietnam, bypassing the northern Chinese ports, before picking up fowl from the northern point of what is now Sumatra.

Compared to local breeds these Asiatic Cochin-Chinas, as they became known, were heavy-boned and clothed in fluffy feathers. They were completely different in dimensions and appearance from any poultry known or seen in Europe. Exhibition of these fowl created massive public interest on an unprecedented scale. In fact, by the time the breed came to general notice at the Birmingham Poultry Show of 1850, the public had already displayed a keen interest in everything Chinese. Writing in 1880 in his *Illustrated Book of Poultry*, Lewis Wright captured the scene and public mood. "Every visitor went home to tell of these wonderful fowls, which were as big as ostriches, and roared like lions while as gentle as lambs: which could be kept anywhere including a garret and took to petting like pet cats. Others crowded in to see them and the excitement grew, and even the streets outside the show were crammed."

▲ *Poultry keeping is thriving, driven by concerns about ethical food. Fowl are a welcome addition to the garden for many.*

▼ *The Asiatic birds were feathery, and had a far greater body size than poultry known in the West, as well as tiny wings positioned high up on their backs.*

Later developments

Led by much hype and publicity, there followed a popular interest in poultry keeping on both sides of the Atlantic. American entrepreneurs started breeding from and crossing the various types of fowl that could be imported either via England or directly from the Far East.

Brahma fowl arrived in Britain from 1852 at the height of the "Cochin boom" when astute breeder and dealer George P. Burnham sent nine birds to Queen Victoria. The case holding the birds was painted purple and gold and addressed to Queen Victoria. *The Illustrated London News* reported their arrival and described them as Grey Shanghais from China.

The imported birds introduced a host of new and useful genetic traits to the Western world, including winter egg production, and the previously unseen brown egg. The possibility of hybrid vigour was to become an invaluable tool to the poultry breeder, later heralding the creation of the modern hybrid.

BREEDS, STRAINS, VARIETIES AND HYBRIDS

What constitutes a breed? In poultry terms it has come to mean the type, or shape, and collection of genetic features that are linked closely enough to be described as a type. There are more than a hundred pure breeds of poultry in the world, many specific to geographical regions.

The diverse visual appearance and characteristics of poultry, such as the inherent ability to tolerate cold or heat, have partly developed in response to the environment in which each breed originated. They are also the result of selective breeding by farmers. Breeders would have prized different characteristics, such as feather patterns or perhaps productivity of the hen, and would have developed these features in their breeding programs.

While few people would have trouble identifying poultry in general, it is likely that many of us will have seen only a few breeds in our local environment. It is the sheer variety of these birds, their differing temperaments and utility or aesthetic qualities, that appeals to poultry fanciers and exhibitors today.

Pure breeds

Nearly all of the important poultry breeds that we now think of as being "pure" have their origins in the 18th century. Pure breeds have distinct physical and visual characteristics. Later "manmade breeds" developed out of crosses between pure breeds. These incorporated various combinations of heavy-boned Asiatic fowl and the tiny ancient fowl of Northern Europe or the

▶ *In some long-tailed Japanese varieties, feather length can extend to several metres.*

Mediterranean in the breeding pen. As a result, modern breeds are capable of much variation. These variations will continue to evolve with the environment in which poultry live as well as from human intervention.

Bantam versions have often been developed in parallel with the exhibition strains. These now often out-number their full-size counterparts.

Few pure breeds rival the egg production of their modern hybrid counterparts, and some are kept to add beauty and completeness to gardens rather than for their output of eggs. It is wonderful that, albeit in some breeds in a miniature or bantam form, so many of the traditional breeds are still here to be studied, understood and reared.

It is these pure breeds that attract many people who see keeping poultry as a fulfilling hobby, an added bonus being a supply of fresh eggs. Today, most hobby or specialist breeders are attracted to a particular breed for its standardized show points, such as intricate feather patterns, modified silky or frizzled

▲ *Feathered feet are generally an Asian characteristic, which can vary from the outer shank feathering to fully feathered feet, where feathering extends to the middle toe.*

feathering, great height and reach or very short legs.

The term "light breed" refers to bone structure and body shape. It is derived from an earlier "sitter" and "non-sitter" classification (meaning those likely or unlikely to go broody and rear their own chicks).

Strains

Different strains of breeds also exist. A strain of a pure breed is one that has been developed by a breeder's family for generations, and has been reared and selected from a closed flock. A closed flock does not allow for any other poultry, including any of the same breed, to be introduced into the flock. In this way the bloodline remains pure through the generations, and the ancestry of the birds can be clearly traced. A strain of poultry may also refer to specific characteristics that a breeder has developed within

▶ *Modified feathering in the Naked Neck breed can extend to a complete absence of feathers on the neck.*

◀ *True bantams, the dwarves of the poultry kingdom, often have a low wing carriage and a jaunty stance.*

his own flock. To the untrained eye all strains of the same breed may look identical. After all, they all must adhere to a written standard for that breed that has been defined and approved by an officially recognized body. Strains may, however, have subtle variations such as egg-producing capacities, or some strains may consume more grass than others.

Varieties

The term "variety" is usually reserved for colour variations within a breed. There are many colours of poultry, a number of which are from selective

breeding. Historically, breeds may have existed in just a few colourways, or combinations of colours. In order to introduce new colours to one breed, poultry keepers include genetic material from other breeds with the requisite shades. Colours such as Wheaten, Buff, Columbian and Silver, for example, have specific breed requirements. Birds may feature more than one colour, and the breed standard may require that a colour be specific to certain parts of the bird, or predominate in a given area. Varieties are also thus standardized, and criteria must be accepted by a relevant body.

In breeds such as Poland, variety is defined by beard or lack of beard. The standard also takes into account different feather structures. When buying imported breeds and new colour varieties with the intention of selling future offspring, check first with an accredited poultry club that the characteristics are to standard.

Hybrids

Often the birds that we see on poultry farms are known as "hybrids" rather than as pure breeds. These are fowl that are in effect artificially bred. They have been developed in response to market pressures to produce the

maximum quantity of eggs for the smallest amount of feed. It was enthusiasts, rather than commercial breeders, who began experimenting with the creation of hybrids more than a hundred years ago. Through observation they determined which hens laid the largest number of eggs, and bred from those individuals. All hybrids have pure breeds in their ancestry. As the industry progressed, hybrids were created by selecting desirable features from different breeds and adding them to the gene pool. Only in a few instances do these hybrids bear any resemblance to the earlier standard-bred utility flocks. Such birds may lay 20 per cent more eggs than their pure-bred counterparts, and are ideal for many domestic situations.

▼ *These hens are typical of those hybrids that lay brown eggs.*

▼ *Muffs and beards often inhibit wattle development.*

▼ *Very short legs are the result of a creeper gene.*

COMBS

It is the comb that distinguishes the genus *Gallus* from other bird breeds. Since the domestication of the ancestors of modern domestic fowl (all of which had combs) there have been a number of mutations affecting the visual appearance of combs.

There are a number of distinct comb types – buttercup, horned, mulberry or walnut, pea, rose, single and strawberry.

Buttercup comb

The buttercup comb is specific to the Buttercup breed of poultry. It is a fleshy comb shaped like a goblet, that sits centrally on the head and is smooth in texture.

Horned combs

V-shaped or horned combs are specific to some European breeds. The two prongs of the V are joined at the base of the comb which starts at the top of the beak. The Houdan, La Fleche, Sultan and Polish breeds all have horned combs.

Mulberry or walnut combs

The mulberry or walnut comb is small, broad and relatively flat, and sits low on the front of the head. It is

▼ *The buttercup comb is unique to the Sicilian Buttercup breed. It should look like an upturned buttercup flower.*

▲ *The mulberry or walnut comb of the Silkie is the shape and colour of the fruit.*

relatively smooth on all sides. Silkies and Yokohama breeds exhibit walnut combs.

Pea combs

The short pea comb is standardized in just five true poultry breeds, including the Sumatra and the Indian or Cornish

▼ *The horned comb of La Fleche accompanies cavernous nostrils, and occasionally a few raised feathers.*

breeds. It was first described in 1850 as being similar to a pea blossom and is a medium-length comb that starts at the top of the beak and finishes at the front of the head. Each comb has three lengthwise serrations.

Rose combs

The rose comb is clearly identifiable by its leader, a spike at the end of the comb that may, depending upon

DOMINANCE IN COMB TYPES

Much of the work into the genetics of comb types was carried out by William Bateson in 1902. He crossed a Wyandotte, which has a rose comb, with a Leghorn, with a single comb, and a Brahma, which has a pea comb. All the offspring (the F1 generation) had walnut combs, an entirely new comb type. When the F1 generation were mated, the resulting F2 generation showed a 3:1 segregation, which was said to have shown that pea and rose combs were dominant over the single combs, thus proving that Mendel's laws of genetics extended to the animal kingdom as well as plants. However, Bateson had chosen Wyandottes, which were the result of earlier unions between breeds with different combs, and this fact could have skewed the results. Comb genetics can be confusing, but fascinate exhibitors and hobbyists, some of whom make the occasional experimental cross. The fact that the purest bred Wyandottes continue to produce a percentage of single-combed offspring could again be the result of their own mixed ancestry.

▲ *Pea combs were thought to resemble the shape of an open pod of peas, or that of the pea flower.*

▲ *A rose comb may terminate in a leader (spike) that extends straight back from a wide front, or may follow the neck line.*

▲ *Single combs with even serrations are expected to stand upright in both sexes, in heavy breeds such as the Sussex.*

breed, be long, short or almost horizontal. The rest of the comb contains fleshy nodules. Several well-known breeds have a rose comb including Leghorns, Dorkings, Hamburgs and Derbyshire Redcaps.

Single combs
Most of the economically significant domestic breeds have single combs, as have *G. gallus*, *G sonneratii* and *G. lafayettii*. With the exception of the Javanese jungle fowl, all the probable

ancestors of domestic fowl have single combs. The large single comb should stand boldly upright in an exhibition rooster, while the comb of its female counterpart may flop gracefully to one side, or may also stand upright. When viewed from the side, the single comb forms a semi-circular head ornament that begins at the top of the beak and travels centrally over the top of the head, finishing at the back. Most combs have serrations, with those at the centre of the comb

standing taller than those at each end. In order to stand upright, the tall rooster comb needs a wide and strong base. To flop to one side the female comb needs to be relatively slim.

Strawberry combs
As its name suggests, the strawberry comb has the appearance of half a strawberry sitting on top of the head at the top of the nose. The Russian Orloff, for instance, has a strawberry comb.

▼ *In some breeds, the single comb should flop gracefully to one side on the hen bird, while standing upright in the rooster.*

▼ *The strawberry comb of the Malay may result from the breed's development in Cornwall, not its Malaysian ancestry.*

▼ *The Russian Orloff breed has a strawberry comb, which starts directly above the beak.*

FEATHER STRUCTURE

Feathers are unique to birds. They act primarily as a form of insulation for the fowl, keeping it warm in cold weather. They also protect birds from the harmful effects of the sun's rays. Feathers allow birds to fly, and also provide a means of attracting a mate.

Apart from in the downy stage, all feathers, including those of adults, can be categorized into three types: short, downy feathers, long feathers and contour feathers. Additionally, the roosters of some breeds may have long sickle feathers at the tail. All breeds have wing feathers.

Down feathers

Young chicks have very fine down, which is quickly replaced by intermediate chick feathers and then by juvenile feathers. Short, downy feathers are found on the abdomen and provide warmth for the bird.

Long feathers

These feathers resemble hairs; they lack barbules except at the tip.

Contour feathers

These cover the wing and tail. Contour feathers include hackle feathers, such as the long feathers covering the neck and saddle.

▲ *The contour feathers of the Silkie breed have delicate shafts and unusually long barbs. The barbules are elongated and are arranged irregularly, not all in one plane as in normal feathers.*

▲ *Frizzled feathers curl in the opposite direction to standard feathers. One form has an extremely recurved shaft and very narrow feathers. Modified feathers are less recurved but have a normal width vane.*

It is the character and distribution of the contour feathers that create the visual outline of the breed. The contour feather comprises a well-developed central quill or shaft, known as the *rachis*. This has barbs, like small veins, that branch off from

it at an angle. The web of the feather is made up of tiny barbules, which radiate from the barbs. These barbules lock together to create a smooth plane, and allow for flight in birds. The lower part of the quill, which attaches to the bird, lacks barbs and is termed the *calamus*.

▼ *Standard feathers, including the quill, silkie and hackle, all have a quill, shaft, vanes and webs.*

Web

Barb

Shaft

Vane

Barbule

Silkie feather

Hackle feather

Sickle feather

FEATHER PATTERNS

Like feather colours, feather patterns are an enhancement that breeders like to perfect, with points on offer for excellence at exhibition. There are less patterns than colours, but being able to distinguish between some patterns and colours can be difficult for the beginner.

Feathers are the most easily distinguishable features on any bird In poultry, this patterning can show in barring, a feather pattern in which two different colours form bands or bars across the shaft of the feathers, or in mottled feathers which are black feathers with a white tip. There may be other feather colours distributed over its body. Laced feathers have two distinct colours, with one of them forming a clear border around the outer edge of the feather. Gold- and silver-laced feathers are the most common. A double-laced feather is harder to perfect since it has two even bands of concentric lacing around the edge of the feather. Pencilled feathers

have a bicoloured concentric band around the feather edge. Like mottled feathers, speckled feathers have a white tip, but with an additional black bar beneath. Spangled feathers

have a blotch of a secondary colour at the end of the feathers. Finally, splashed feathers are a dilute form of blue and are not a standardized colour.

▲ *Barred*

▲ *Mottled*

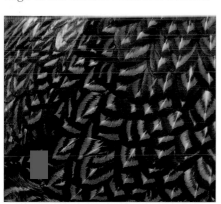

▲ *Double laced*

▲ *Pencilled (female partridge varieties)*

▲ *Speckled*

▲ *Laced*

▲ *Spangled*

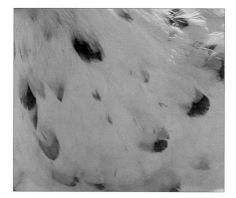

▲ *Splashed*

FEATHER COLOURS

The colour of a bird is a significant visual feature. Breeders who choose to exhibit their prized poultry may spend a lifetime dedicated to perfecting the feather colours of their breed, since breed standards are exacting and marks will be deducted for wrongly placed colours.

New poultry colours are constantly being developed, although it takes time for them to be accepted by the ruling poultry club for each country.

Like the feather patterns, feather colours can vary between roosters and females of the same variety. As exhibition colours evolved over

hundreds of years, there is no logical format to the colours of roosters and hens, and they have to be learned variety by variety.

▲ *Birchen (as seen in some Oxford Game)*

▲ *Buff*

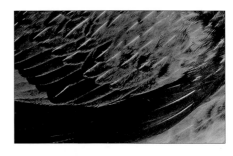

▲ *Furness (female Old English Game)*

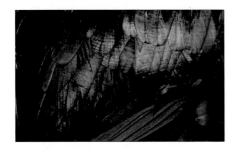

▲ *Black*

▲ *Crele*

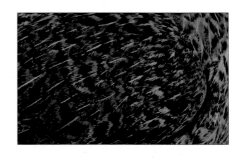

▲ *Ginger (female)*

▲ *Black red (male)*

▲ *Cuckoo*

▲ *Gold*

▲ *Brown (Leghorn female)*

▲ *Dark (Indian Game)*

▲ *Jubilee (Indian Game)*

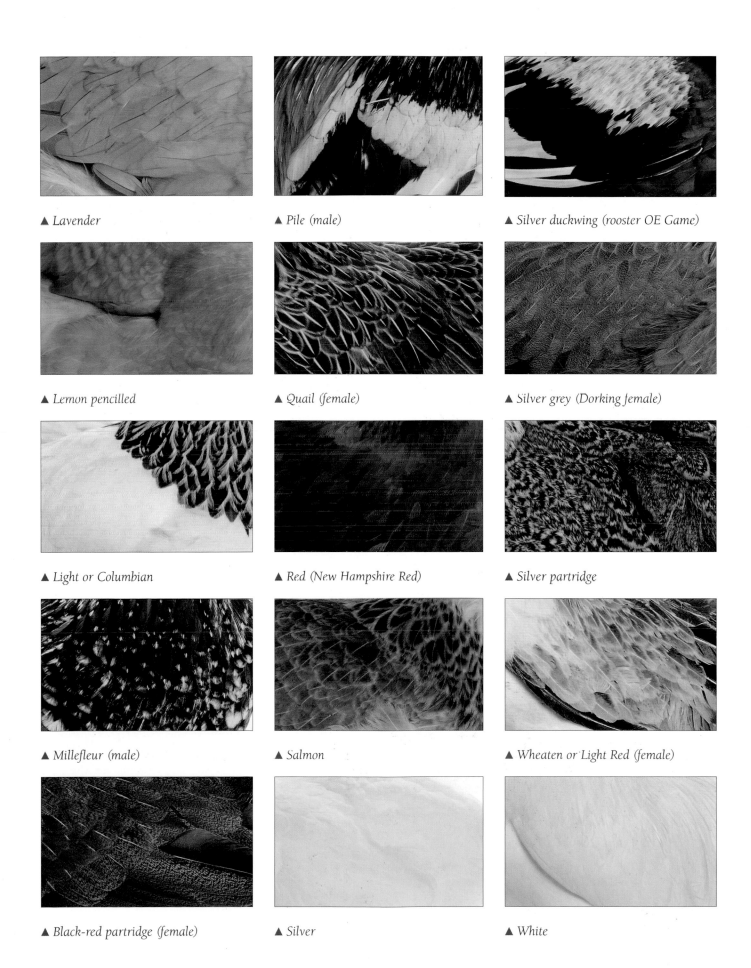

▲ Lavender

▲ Pile (male)

▲ Silver duckwing (rooster OE Game)

▲ Lemon pencilled

▲ Quail (female)

▲ Silver grey (Dorking female)

▲ Light or Columbian

▲ Red (New Hampshire Red)

▲ Silver partridge

▲ Millefleur (male)

▲ Salmon

▲ Wheaten or Light Red (female)

▲ Black-red partridge (female)

▲ Silver

▲ White

THE LIFE CYCLE OF POULTRY

On average, poultry live between three and five years, with some breeds living longer. They have three distinct life phases: it takes approximately 25 hours for a hen to form an egg. A fertilized egg takes an average of 21 days to hatch. The hatched chick takes six months to mature.

A broody hen will lay a clutch of eggs before she sits on them: up to 12 eggs, with each egg being laid a day apart. Only those eggs fertilized by the rooster will hatch. The hen sits on the eggs in order to maintain the correct temperature for the chicks to develop, turning the eggs every day until they are almost ready to hatch.

The fertilized egg is a capsule of nutrients for the developing embryo. It contains the white, the yolk, an air sac and most importantly a blood spot, which is the beginning of new life. While it is growing in the egg the chick takes its nutrients from the yolk. As soon as the chick hatches, it will take very little time to dry off and stand up.

Birds develop their features quickly. The cute and fluffy chick stage lasts for approximately a month, at the end of which the bird will be developing

▼ *A broody hen rarely leaves her eggs except to take care of her own needs.*

▲ Running a rooster with the hens is only essential if you want fertilized eggs.

the characteristics that identify it with its breed. Within six months the bird will have all its adult features.

▼ *The newly hatched chick can walk almost as soon as it breaks out of the shell.*

▲ A broody hen exhibits specific features and behaviour, such as a raised body temperature and sitting for lengths of time.

Young roosters
Many young rooster chicks are discarded as soon as they can be sexed, particularly if they are bred for egg production. Some may be reared for meat, or to produce the next generations. Roosters develop quickly, with some light-breed bantams mature enough to sire chicks at four months olds. Most large-breed roosters are at their most virile between six and 18 months old.

The developing hen
Some females may be ready to lay at 18 weeks, although this will vary between breeds; young hens are known as pullets. All birds tend to come into lay earlier during periods of increasing daylight. For instance, productive hybrids that normally start to lay at 18 weeks may start laying tiny eggs as early as 14 weeks, if the birds reach maturity about midway

▲ *Chicks grow and develop quickly. The first true feathers can start to appear in the wings as early as three days old.*

▲ *In this pair of seven-week-old growers, the rooster (right) can be identified by his developing comb and wattles.*

▲ *All birds have an annual moult, during which time they do not lay eggs and start to look decidedly scruffy.*

through spring. Some exhibition-bred pullets may delay laying their first egg until they are well over six months old if they reach maturity in early winter. Productive females will lay eggs until 72 weeks old, depending on the period of their first moult.

▼ *This ten-month-old heavy Sussex breed is in the breeding prime of his life.*

Those in full lay by high summer of year one will often lay right through the winter, moulting in midsummer of the following year.

Ageing hens

Birds have an annual moult, which generally occurs in late summer or early autumn. At this time a bird will lose its old feathers in a specific order working from the head and neck down the body. The result is a bird with a tatty appearance as if in a state of undress. While moulting, birds cease egg production and all energy is directed to providing the bird with a new covering of feathers. While some birds will quickly recover their plumage and soon start laying again, many of those hybrid hens that have already laid close to 300 eggs may be expended, and it is standard commercial practice to dispose of them at this point. Yet given sympathetic treatment in a home environment, they may in fact, continue to lay a similar weight of eggs in the following year.

Some hens, if they live long enough, may continue to produce a dwindling number of eggs for years. Poultry may live for ten years or more.

▼ *A wrinkled and dull face, heavy eyebrows, a sagging underline and an increasing rear width are among the many possible indications of approaching old age in poultry.*

BUYING POULTRY

There are more than a hundred different pure breeds of poultry, including many that you have probably never seen or even heard of, and all have unique characteristics. As well as pure breeds, hybrid poultry are available. Hybrids have been bred for maximum utility value and make up a significant proportion of domestically kept poultry. Being clear about why you want to keep poultry will help you decide whether to keep pure breeds or hybrids. You may like to own hybrid hens if your requirement is for large quantities of eggs or for poultry meat. If an unusual pet appeals, as well as the chance to have a few eggs, then a pure breed may be what you're looking for. Such breeds also bring a rewarding pastime, and those keen enough to breed and rear their own chicks may take pleasure in selecting the best and exhibiting them at a local show. Whatever your reasons for keeping poultry, take care to choose the correct breed for the area in which you live, the amount of space you have available, and the amount of time you wish to devote to poultry keeping. This will ensure that the transition to ownership is as easy as possible. Always purchase your birds from a reputable supplier.

▲ *Hybrid parent stock are nearly always kept in bio-secure surroundings. This should result in disease-free stock.*

◀ *A knowledgeable breeder will be able to point out genetic complexities, in this case, that the black ticking seen in these mature Red Sussex breeding hens is due to their carrying the gene for slate undercolour – an important feature of the variety.*

DECIDING HOW TO BUY

For those new to poultry keeping, beware – the hobby can become addictive. Often, people who buy chickens for the backyard start with just a few birds (three is a good number). They then gradually increase their stock as their confidence grows.

Poultry can be bought at different life stages, including as eggs to hatch, as fully feathered growers and as point-of-lay pullets.

What age to buy?

If you would like to see poultry hatching and be responsible for your poultry from when they are chicks, it is possible to buy fertile eggs from reputable breeders. The advantage of purchasing eggs is that you can question the seller about the antecedents of the parent stock, and how it has been reared. It also means that the poultry is reared according to your own standards after hatching.

The disadvantage is that, for those who want hens only, any one batch of eggs is likely to yield 50 per cent females and 50 per cent males, so unless you want to rear roosters, are able to rehome them, or want the

▼ *Poultry auctions may be a good way of replacing stock if you are an experienced poultry keeper.*

▶ *Experienced auction buyers often take a long look at intended purchases in an attempt to spot possible defects.*

responsibility of culling them, this might not be the most desirable option. There is also no guarantee that all the eggs will hatch, and if they don't it may leave you with less stock than you might like to have. Buying hatching eggs is probably only a viable option for those with some experience of poultry keeping.

Chicks need more intensive rearing than hens, and may need different housing and feed, so unless you are able to provide these, then buying hens at a later life stage may be more appropriate for your needs. It is most common to buy hens as they are approaching their point of lay. Most hens start to lay eggs at around 18 weeks of age. This is the optimum time to purchase hens, since any feed that you buy is converted almost immediately into eggs. Poultry

can be purchased at a younger age, but a reputable breeder is only likely to allow you to have them if you have both the facilities and the knowledge to be able to look after them.

Where to buy?

Breeders may advertise their stock in poultry magazines, on the internet, in a local paper if you live in the country, or, if their reputation is good, by word of mouth. Poultry fairs run by local clubs are an alternative way of purchasing stock, or better still, of gaining an introduction to knowledgeable breeders. If you are looking for one of the rarer breeds,

▼ *As ducks do not suffer from the same endemic respiratory infections as hens, they may make safer auction purchases.*

A PRELIMINARY HEALTH CHECK

Check any poultry thoroughly before purchasing it. Look for areas of broken or missing feathers, particularly on the lower back, as this can indicate a degree of overcrowding in the pen, and can lead to later problems with feather pecking. If it is possible to handle the birds, then you always should.

A round, clear eye is a good indicator of health. Any bird that shows any sign of nasal discharge should be avoided. If several birds or pens are affected, it could indicate a contagious infection is present. A clean vent indicates a healthy digestive system. Dried faeces could indicate an infestation of worms, while clusters of lice or lice eggs in this area show that there is an overall health problem. If these are present, the whole purchase is best abandoned.

▶ *A healthy grower of the relatively rare, blue egg-laying, auto-sexing Crested or Cream Legbar breed.*

then you may have to join a breeder's waiting list for stock.

Impulse buying, often from unregulated markets, is an unsatisfactory way to begin poultry keeping. Poultry auctions give little chance of making any real assessment of birds' health or background, especially if you are a beginner. Some of the birds for sale could already have been purchased at auction. For those wanting to keep a few hens at home, poultry auctions are best avoided.

Things to bear in mind

Keeping poultry is a commitment made for the lifespan of the poultry. Whatever the weather, poultry need daily attention, fresh food and water, which makes going away tricky, unless you have friends, neighbours or relatives willing to help out. The birds also need cleaning out and their coop to be maintained. Routine handling must be carried out to check each bird for health problems. Should a bird become unwell, its medication or veterinary care will need to be paid for.

Some breeds are more expensive to keep in food than others. Some like a free-range life, while large breeds will need bigger and more expensive housing units that take longer to clean. You also need a sizeable area in the garden in which to keep them. The longer one keeps poultry, the clearer it becomes that every aspect of their management is interdependent. Take time to think through every aspect of owning and keeping fowl before making a purchase.

Pure breed or hybrid?

If you are clear about why you want to keep poultry, then you will soon know what type of poultry is suitable for your situation. For those whose primary interest is in having fresh eggs, a modern hybrid may best fit your needs. So large is the gap between the numbers of eggs that hybrids lay, compared to the dozen most productive pure breeds, that it is probably best to look at the opportunities for purchasing hybrid fowl first. For those who want a family pet, a hybrid is also suitable, though many pure breeds make perfectly good pets – just choose a breed with a suitable temperament.

▼ *Poultry bought from a breeder can be seen in the surroundings in which it has been reared. A large, airy house and plenty of fresh air provide a good start toward keeping healthy birds.*

CHOOSING HYBRIDS

The hybrid hen, which is now responsible for laying most of the eggs consumed in the developed world, is not a true breed, although hybrids have pure breeds in their distant ancestry. Hybrid hens are the best choice of bird if you are interested in quantities of fresh eggs.

Hybrids were developed in the 1950s in response to the need to supply burgeoning population levels with cheap meat and eggs. Poultry are bred to be efficient producers of large numbers of eggs with the least outlay. The battery system of housing and keeping poultry evolved in response to this demand. Prior to the successful introduction of the hybrid, poultry were reared outdoors, and were prone to disease. Hybrids are now resistant to disease, and are often inoculated to reduce the risk of sickness.

Often people selling or rearing these popular birds will have difficulty in describing their birds' grandparents. In fact, laying fowl sold as hybrids are very like the best of the earlier crosses between high performing in-bred lines of traditional pure breeds.

Egg numbers

All hybrids lay extremely well. They are available in a number of plumage colours, with the various types laying

◄ Hybrids can be expected to lay between 280 and 300 eggs in their first year.

eggs of a slightly different hue. Nearly all of the brown egg-laying hybrids are incredibly quiet, and batches of these bought directly from a reputable source and observed and handled for a year from point of lay should provide the perfect benchmark to judge the performance of other

groups of fowl. The paler brown poultry with nearly white tail feathers are used in battery cages as well as low-density, free-range systems and have proved to be very successful backyard birds.

In performance terms, there is little to distinguish between the main competing hybrid lines supplied to the egg-producing industry. All hybrids are reared within the bio-security of a housed system. All are vaccinated against most of the troublesome respiratory diseases that infect hens, and the salmonella strains that can infect eggs intended for human consumption. These hybrids will usually leave their rearing unit at 14 weeks old and are supplied to farms in large batches.

If you are fortunate, you may be able to share a batch with another customer. The more usual option, however, is to buy from a trader who takes pullets from a hatchery or grower (the point of origin) and has reared them on, ideally in isolation from other indiscriminately purchased fowl. It should be possible to see how they have been reared and to ask about feeding regimes, vaccinations, health issues and on-going care. All hybrids from such a source are expected to have a good health status, but always check.

◄ Those hybrids used by the industry will be bred in small units, and are best purchased as close to your home as possible. All should have good health status.

CHOOSING PURE BREEDS

There are more than a hundred pure breeds of poultry to choose from, all with different characteristics, temperaments and requirements. Pure breeds are generally less productive than hybrids, but many are unusual, with interesting features and attractive feather markings.

Owning a pure breed of poultry is extremely appealing to many people. Pure breeds are less productive than hybrids, though egg numbers vary according to the breed chosen. Pure heavy breeds may lay between 60 and 180 eggs per year; light breeds may lay more. If eggs are a secondary consideration, the pleasure and satisfaction of keeping one of the old and attractive pure breeds will outweigh the egg numbers or the cost of their upkeep. Occasionally, a utility strain of pure breed capable of laying 240 eggs a year may be found. Pure breeds are usually more expensive to keep than hybrids because they eat more food and lay fewer eggs, making the unit cost of eggs more expensive.

With few important exceptions, all birds bred to conform to a breed standard will have visible characteristics that have been influenced by exhibition results. All standard-bred fowl should be capable of breeding similar offspring to

▲ *A blue-mottled Pekin makes an excellent garden companion and broody hen, but is likely to be a poor egg layer.*

themselves, which can, with selective breeding, provide future generations of useful and beautiful fowl.

Pure-bred poultry may also require larger housing and more space to accommodate their size, unless, of course, you choose bantam pure breeds. Many of these miniature fowl are about a quarter of the size of the original breed but replicate them in looks and character. In many circumstances, these make ideal garden fowl. Strains of miniature heavy breeds such as the New Hampshire Red lay plenty of reasonably sized brown eggs. Poultry keepers with the patience to handle and quieten the excitable miniature

▶ *When purchasing a table-type bird (for eating), one would look for both an active fowl and a well-developed breast.*

Mediterranean breeds are likely to be rewarded with quantities of surprisingly large eggs.

With countless beautiful breeds to choose from, it is understandable that many beginners wish to buy pure breeds rather than hybrids.

While vaccination may offer most of the commercially reared hybrids some level of protection, those purchasing and breeding pure-bred stock will generally have to rely on the natural immunity of the birds they have chosen. If the bigger commercial outlets are good at rearing hybrids, the reverse is probably true when it comes to the less common pure breeds. Often, the best stock is found in the hands of hobby breeders. It may be that once you have located a source of your chosen breed, you have to order stock for the coming season and wait for its arrival. Established breeders are often willing to help newcomers on the road to becoming competent poultry keepers.

CHOOSING RARE BREEDS

The term "rare" can be misleading when applied to poultry. In some countries, the term implies that a breed is endangered in some way and that few of its kind exist. In other regions, it merely denotes a breed that is not popular enough to possess its own breed club.

In some countries the term "rare" applies to every breed that does not have its own breed club. This is based on the premise that there are not enough breeders to meet on a regular basis to promote a club for the breed.

The wish to conserve a breed that is ancient, beautiful or rare may be an important factor in the decision to keep specific poultry breeds. Rare breeds are seen less regularly on the show circuit. They may not necessarily be any more expensive to purchase, however, than breeds deemed to be less rare. Many of these breeds remain rare, not because they have fallen out of fashion, but because they are too flighty to be

exhibited easily. Others may be such poor layers that they do not readily reproduce. Alternatively, having been rare for so long, the existing stock may be so inbred that the birds have health issues, and stamina has become a problem. The enthusiast breeder may also be far more

▲ *A red-saddled Yokohama rooster and blue Sumatra pullet enjoy freedom to range. Such an environment helps to promote a healthy constitution.*

interested in the prospective purchaser's ability to keep and maintain the breed.

▼ *As the Appenzellar Spitzhauben is a productive and fertile fowl, it need not remain rare, but few people will be able to give it the extensive range that it enjoys; others will find it difficult to handle this extremely active fowl.*

LOCAL POULTRY CLUBS

Clubs vary in the way they operate, but are a good starting point in the search for birds, breeders and information. If you are new to poultry keeping, joining a club will provide you with a point of contact. Some clubs organize an annual show, while others see their role as supporting poultry keepers. They may have quarterly or monthly meetings, and arrange guest speakers who will cover all aspects of poultry keeping. The local club secretary will be able to answer most beginners' questions, or may know a member who can.

RETAINING THE INTEGRITY OF THE BREED

Prior to the development of modern egg-laying hybrids in the 1950s, pure breeds were developed for their utility value as well as for exhibition purposes. Utility breeders wanted hens that were productive, laying the maximum yield of eggs for the minimum amount of outlay. Heavy breeds of poultry, which would yield a large carcass, were also developed for the table. Following the introduction of the hybrid hen, many utility strains died out as breeders replaced their pure-breed flocks with the potentially more lucrative hybrids.

Today's pure breeds are predominantly exhibition strains. In order to develop these birds to accurately match the written standard for the breed, breeders have selected and developed their strains on the basis of their visible characteristics, such as shape, leg feathers and plumage patterns, at the expense of the utility value of the breed.

The visual appearance and characteristics of many breeds can easily be recreated, providing the required features exist in other breeds. The Sultan breed, for example, was at one time lost until "remade" by crossing other existing breeds that could donate one or more of its features, such as its head crest or vulture hocks. Extreme care needs to be taken to ensure that, whether utility or exhibition, each breed is maintained to a high standard so that mediocrity does not become the norm. Mediocre birds look like the breed type but do not perform as such.

Because breeders have introduced genetic material from different poultry breeds into the breeding pool of specific

▲ *The Marsh Daisy is an English breed developed in the early 20th century. Fifty years later it was assumed to be extinct, until a flock was discovered. It is now endangered.*

breeds in an attempt to increase the plumage colours, for example, or to improve some other physical characteristic, it is likely that many breeds will lose much of their unique genetic imprint.

At one point, thousands of birds were developed from one highly productive family. The degree of inbreeding needed to fix desirable features within the breed meant that each strain developed its own genetic pattern (genotype). As long as there were thousands of similar flocks, strains of each breed could continue to evolve when crossed with similar genetic lines. Strains kept in huge numbers allowed for some form of reciprocal flock mating so that they could survive as a closed flock.

Huge changes in industrial poultry keeping have seen nearly all of these important flocks virtually disappear. Luckily, some individual breeders have bucked the trend and kept flocks in their families for generations.

▼ *The White-faced Black Spanish breed is difficult to perfect with its all-white mask-like face.*

REHOMING BATTERY HENS

Battery hens are fowl that have been intensively housed for the purposes of egg-laying by the poultry industry. They no longer have a value to the industry when they reach a point where they are deemed uneconomic to retain. Rehoming battery hens is a moral concern for many.

Intensively farmed poultry has received bad press in recent years. Yet for more than half a century, battery hens have produced eggs and provided meat to feed the world, and will continue to do so in the future.

Historically, poultry was kept along with a range of other livestock on a small-scale home farm. As egg and meat production were given over to intensive farming methods providing sufficient food to feed the world's growing population, the nature of poultry keeping changed. Large-scale, totally enclosed buildings became home to hybrid birds specially bred to produce large quantities of eggs for very little feed. Such housing units can hold a minimum of 10,000 birds, kept in an artificially mantained environment where food is constantly available, and artificial light is provided for 17 hours per day in order to maximize egg production. The intensively farmed hen, guaranteed to lay an egg a day, was condemned to spend its adult life in a deep litter system, unable to express

▼ *Battery hens are kept in cages with a continual supply of food in front of them. They have little room to move.*

its natural instincts and inclination to run, take a dust bath or even stretch. After a time, it was found that hens could be housed in wire "battery" cages suspended above each other. Each poultry house could now accommodate larger numbers of birds in the same space, resulting in a greater yield of eggs.

Such birds are subject to infestations and disease, and often resort to feather pecking and cannibalism to relieve their stress. Revised legislation is due, which will alter the housing conditions for hens, but currently it is legally acceptable to house nine hybrid hens in a cage that measures 1m/1yd square, providing each hen with the floor space of less than an A4 sheet of paper – less than the hen needs to be able to flap its wings. Millions of fowl are still kept in conditions that are now considered inhumane and cruel. Eggs produced by such hens are likely to be those that are the cheapest to purchase.

▲ *Welfare organizations work with farmers to relieve them of stock that are no longer deemed productive. Such birds are found new homes, and make good garden pets.*

Consumer power

With growing public awareness and concern for the welfare of the birds, as well as for the food chain, consumers and campaigners have begun to demand more humane treatment for hens. In consequence the market for free-range eggs has increased. Conditions have improved for battery hens; legislation means that farmers of intensively housed birds must now provide cages with perches and dust baths so that hens can take the opportunity to move more freely and take a dust bath. As a result, most new poultry units are now able to describe their hens as free-range, but even the hens kept in these systems will still be considered expendable after laying close to 300 eggs.

▲ *An intensively farmed hen, with feathers missing but in relatively good physical condition.*

▲ *The same hen after just a few weeks in new surroundings. Ex-battery hens remain productive once rehomed.*

ways that are instinctive to free-range fowl. About 20 per cent of rehomed battery hens will not survive the ordeal of rehoming, despite the fact that they are moving to better conditions. Generally, however, rehomed birds will take very little time to adjust to their new lifestyle, usually just a matter of a few weeks, during which time they quickly grow new feathers and start acting according to their natural instincts.

Productive lifespan

Hens start to lay eggs at approximately 18 weeks old. At this life stage, hybrid hens will routinely lay one egg per day and continue to do so until they reach 72 weeks. Having laid an enormous amount of eggs in their first season, the hens need a recovery period. From this point on they become a burden to the intensive factory production method, since the cost per egg increases as the hens lay fewer but continue to eat the same amount. Such hens are deemed to be beyond their useful life and are treated as scrap by the industry. Most commercial units replace these poultry with a new batch of younger fowl.

Rehoming

Given a short rest period and sensible management ex-battery birds will continue to lay eggs for at least

▶ *Rehomed hens will enjoy rooting in deep straw, especially during the long, dark winter months. All birds are often less productive when daylight is short.*

another year, though not at the same rate. These hens are sometimes sold on to farmers who have adequate, but less costly, housing.

Rehoming battery hens has become a popular and rewarding method of acquiring stock at very little cost, and there are plenty of organizations that specialize in finding new homes for hens that have been rescued from almost certain death. Such hens are likely to be in an unhealthy condition, with feathers missing, and the sight of such birds may be distressing. Many hens are unused to having freedom to roam, and are unable to behave in

UNDERSTANDING THE TERMINOLOGY

Free range can mean complete freedom to range, but when stock is confined, it is used to define a stocking rate of fewer than 80 birds to the acre. Most eggs sold as free range are produced by birds housed in large, well-designed houses that allow birds unrestricted access to an outside run. This system would once have been described as **semi-intensive**. **Organic** refers to the means of producing the birds' feed, but the Soil Association also includes very strict welfare and stocking-rate criteria. **Battery housing** allows four or five birds to be housed in one tiny pen. The new **enhanced battery pen systems** allow the occupants to perch, stretch their legs and take a dust bath. **Intensive housing** systems may include large, barn-like structures that allow plenty of room for the occupants to stretch and take a dust bath without outdoor access. Some **deep-litter systems** allow straw and litter to build up, which is often only removed on an annual basis. Deep straw in an open-sided yard or lean-to can add much to the welfare of free-range poultry during winter and long wet spells.

It is standard industry practice that all hybrid birds kept in larger units are rarely retained for a second year.

HOUSING YOUR POULTRY

Poultry housing must provide shelter from wind, rain, sun and excess heat. It is a place that provides protection from animal predators, yet contains enough room to move and follow their natural behaviour patterns. Fowl have a respiratory system that is intolerant of poor ventilation, therefore the housing should have fresh air freely circulating by adequate ventilation, but at the same time, be draught-free. A roost for the birds to perch on and sleep in comfort at night is essential, as is a nest box in which to lay eggs. Night-time accommodation should be dark, with no windows. The environment should be kept clean, with a fresh supply of drinking water, grit and food, and with an access route to daylight. For many breeds, there should ideally be the opportunity to spend time in an outdoor environment as well as to eat fresh vegetation. Additional access for the owner to inspect the internal house and the birds, as well as to collect the eggs, is essential. Outside the house, the birds require a dust bath in which to clean their feathers of fleas and lice, though many birds will create a space of their own in the garden, if they are able, in which to bathe.

▲ *Fowl have a primitive requirement to perch or roost at least 1 metre/3 feet above floor level.*

◄ *This movable house under trees offers plenty of shade, but the ladder may be too much of an open invitation to the poultry to roost in the trees at night.*

CREATING THE IDEAL ENVIRONMENT

Wild jungle fowl, the distant ancestors of today's domestic poultry, would have lived in arid scrublands, roosting at night in trees. Their "run" would have been in an area with a canopy of trees nearby to avoid aerial attack, and with the option to fly to safety to avoid ground predators.

Domesticated poultry have retained their wild homing instincts. From their point of view, the perfect environment is a small, insect-rich patch of land, containing sufficient organic matter to encourage a succession of manure worms that will provide a ready supply of food. An accompanying overgrown hedgerow in which to shelter, as well as forage for a variety of berries, would constitute poultry heaven. A poultry house in the middle of a bare field, with plenty of land in which to roam, represents a free-ranging environment to poultry. When allowed to roam according to their natural instinct, hens produce good-quality eggs and, at the same time, enhance our environment.

▼ *A carefully trimmed windbreak, just far enough from the hen house, provides shelter and allows free movement of the air through the wire pen fronts.*

Choice of site

Careful thought needs to be given to the siting of a permanent poultry house, more than for one that can be moved with the seasons within a large enclosure. Keeping even a few hens is a commitment, and this may be less appealing in the winter months, when

▲ *A house that is positioned for shade under the tree canopy in the summer can be moved further away from the trees and into the light to benefit the birds during the shorter winter days.*

the weather is inclement and daylight hours are limited. For this reason, choosing a site within easy range of the house makes for convenient poultry husbandry.

Remember to consider your neighbours when planning any new poultry house or venture. A rooster crowing early in the morning is not everyone's idea of a perfect alarm call. However, the gentle clucking of hens going about their business may be more appealing, particularly if the neighbours are offered the chance to share some of the eggs.

A well-maintained hedge, like a tree, can shield the poultry house from any prevailing wind. Trees that provide valuable summer shade can also leave smaller enclosures dark and damp in winter, as well as creating a

▲ *Cockerels are likely to crow from dawn to dusk as well as during moonlit nights. House and street lights can also cause unwelcome night-time crowing.*

roosting place for wild birds whose droppings have been shown to assist the spread of serious poultry diseases.

It may be possible to site a "lean-to" poultry house or an enclosure against a boundary wall or close-boarded fence. To avoid either scorching summer heat or almost total shade, it will have to be carefully sited, with adequate guttering to prevent dripping water from reaching the covered area.

Daylight and climate

Fowl, particularly laying hens, are more light-sensitive than any other form of livestock. They need plenty of daylight and for this reason early laying houses had large windows and were positioned to capture the maximum hours of autumn and winter daylight.

In a temperate climate, protecting fully feathered fowl from rain and damp is likely to be of greater importance than protecting them

from cold. In a wet climate, a covered area (other than that provided for sleeping) in which to feed and shelter is an essential part of good housing. In high rainfall areas, intensively housed birds kept in static houses will benefit from an entirely covered run. In a hot climate, however,

▼ *A sheltered corner away from the main house is always attractive to poultry of any breed.*

◄ *The garden layout should be planned to provide a clean and pleasant environment for the poultry and their carers.*

providing screening to protect birds from intense heat is imperative.

Many breeds will not cope well with very strong sunlight in summer and will need shelter from it. This latter problem is soon remedied by placing the house close to walls, fencing or vegetation and partially covering a section of it so that the poultry can get out of the sunlight, should they need to. The hen house will need to provide sufficient space and ventilation for larger, fully feathered birds during summer nights. Many birds were once housed in an environment that had netting sides and an overhanging apex roof to assure full ventilation. A fully feathered hen is, in effect, wearing a feather duvet, and a batch of ten growing pullets can generate as much heat as a single-bar electric fire.

▼ *Placing the feed dish under cover can prevent both birds and feed becoming sodden during a rainy day.*

CHOOSING APPROPRIATE HOUSING

Choosing poultry housing that is suitable for the number and type of birds that you intend to keep should be your primary consideration when deciding which type of housing to purchase. Ensuring that you have space to accommodate the house and run is also essential.

There are legal requirements for the amount of space that poultry needs although most poultry housing available to the enthusiastic amateur will exceed such prerequisites. An off-the-shelf system may be a wooden or plastic house, with solid or slatted flooring. The outside should be easy to maintain and tanalized against the weather. The roof should overhang the side walls and be angled so that water drips away from the entrance.

Roofing material is significant, since some types allow mites to live in the surface. When covered with felt close-boarded roofs can provide the perfect breeding ground for mites, which are difficult to treat and remove. A good quality plywood base

▼ *Poultry owners planning to move a house on a regular basis should look for a house with wheels or skids for ease of movement. A small garden tractor may be needed to help move larger houses.*

to the roof, clad with corrugated sheets, can minimize this problem, and allows for a well-directed spray of crevice treatment should it be needed.

If the floor is a solid wood or plastic, it provides a warm and comfortable surface for the poultry to

▲ *A well-ventilated run is important, but some shelter from sun and rain should always be available.*

walk on, though is more labour-intensive when it comes to cleaning. Slatted surfaces allow droppings to fall through the flooring, and may be colder to walk on since air and cold circulates from below. Wire mesh floors may be an uncomfortable surface, and should ideally have a tray beneath, from which the droppings can be collected and removed on a daily basis. Solid floors are easier to clean if done regularly and can then be covered with no more than 1cm/½in of clean wood shavings or high-grade white sawdust. Clean hay should not be used other than to line nest boxes.

Space requirements

For those who have only limited space, it is best to determine how much of that space you have available in a suitable position, and to choose

the breed of bird accordingly. For instance, both large, heavy breeds and active breeds will benefit from an additional covered area in which to exercise. Most of the crested breeds such as the Poland are best housed in a roofed enclosure for most of the year, because of their intolerance of wet crests. Some of the "primitive" light breeds like the Appenzellar Spitzhauben, which require space to roam and to exercise their natural instincts, will be quite at home in a tree – predators permitting.

Miniature versions of the popular heavy breeds such as the Orpington appear to be happy in close confinement. Some strains of true bantams value human contact as well as an extensive run, so ensure that you have comfortable access to their hen house. Large birds clearly require more space than small, and active birds need more than docile fowl. Most standard housing can be modified to also house gigantic breeds, if that is your choice, and if your DIY skills are up to it.

Budget or deluxe?

Since most people keep hens for their eggs, much modern poultry housing is designed for hybrid egg-layers.

▲ To accommodate large fowl without the run becoming soiled, this combined house and run has to be moved on a daily basis. The run should be partially covered with translucent material in wet weather.

Certain pure breeds may need more space or differently designed internal features, so bear this in mind when choosing your house. With a huge number of small- to medium-size houses on the market, people thinking of keeping poultry or buying a new poultry house should draw up

a list of requirements that match the breed they wish to keep and the amount of space to hand.

So many types of housing are available that it can be difficult knowing which to choose. Generally, purpose-made poultry housing is the most economical and convenient option. It is unlikely that you could make a basic poultry house as cost-effectively as the cheapest versions on the market. There are any number of types available, catering for all budgets as well as poultry numbers, ranging from deluxe, expensive models through to kit form options. Though "flat-pack" options are convenient, they may be of flimsy construction, and some purchasers question their value for money.

If birds are to be housed in a prominent position in the garden, keeping them in view and a daily reminder of their need for attention, many owners will want the house to be pleasing and complement its setting.

▼ Modern pressure treatment will add years of use to a wood-clad house, but it is worth finding out if the timber framing has also been treated.

TYPES OF HOUSING

Many different poultry-house types are available, including semi-intensive systems, combined house and run forms, and free-standing static houses in which an area of land needs to be penned in for a run. Different sizes, to suit various breeds and bird numbers can be purchased.

Keeping small groups of poultry in a garden situation requires well-planned housing. You will also need a management plan that takes into account wet and cold weather as well as short winter days, when poultry barely have time to consume enough food to last them through 18 hours of darkness.

Semi-intensive systems

For first-time poultry owners who want a few hens to provide eggs for the household, a small-scale semi-intensive housing unit may be a good choice. This is a construction where the poultry are locked into an enclosed area at night, protecting the hens from thieves and from foxes. The birds have space to behave

▼ *This semi-intensive house may not be easy to move. There needs to be 20 times the area of the run available in the garden to accommodate moving the house.*

naturally in an area where they are free to move either below or to the side of the sleeping area. Such housing allows for the highest standards of husbandry and welfare for poultry, since it is small, easy to access and to keep clean. Most semi-intensive units are designed for

▲ *Small henhouses, such as the A-framed ark, should ideally be movable, either with wheels attached beneath them or via strong handles, so that the run area beneath or to the side of the house can be moved on to fresh, untainted grass.*

ultra-productive hybrid hens whose small size is appropriate for the space. Many of the smaller units are only large enough to house miniature fowl, while some only have enough room to move for the smallest true bantam. They are inadequate for larger fowl, particularly those profusely feathered examples with feathered legs or feet.

The upstairs-downstairs ark is a triangular house, in which the occupants roost at night in the apex. There is an enclosed part of the housing at the top of the A frame, offering a comfortable and secure sleeping area. The grass-floor area is usually enclosed in wire mesh, offering daylight and access to natural vegetation. Connecting the upper and lower levels are steps or a ramp that

the poultry must climb; an exercise important for the birds' wellbeing. The occupants can be contained within the housing without the addition of a run, provided the natural flooring does not become too muddy. Once it is, the house should ideally be moved to a new patch of land. This housing design has proved so successful that it has assumed a pivotal place in the market. It provides an opportunity to inspect the birds regularly, particularly in the evening when they are roosting, since it is usually designed to allow human access to the roosting area. At the same time it allows easy access to inspect the inside of the house, which is helpful for detecting troublesome infestations of red mite. Designed for productive hybrids, such an ark may also be used for some of the flightier light breeds.

The drawback to this type of housing is that most modestly sized fowl will end up walking to and fro in a very narrow path in the sleeping area of the ark, where they can stand fully upright with ease. In showery weather, the downstairs area will quickly wear to a muddy track.

Unless the food is covered it may also get wet, so that the unfortunate occupants will, on a really wet day, be faced with a choice of going to bed wet or hungry. There is also some potential for stress bullying in small or intensive housing.

Combined house and run

The combined poultry house and fixed run makes the perfect solution for small-scale poultry keepers. Many square poultry houses are designed with a "pop hole", which is basically a trap door that allows the hens access outside. Many square houses without a run integral to the design have the option of attaching a small run to them, so that the poultry can spend time in daylight and have access to vegetation. The run is an enclosed area usually made of wire mesh, so the poultry are free to move but are not free to roam around the garden if there is no one to keep an eye on them. Some designs have a slatted

▼ *The type of house where the owner can walk in and spend time looking at the birds on their perches has a distinct advantage over many other versions.*

▲ *This free-standing house will require two people to move it to a fresh site.*

floor cleverly sloped in two directions to give the birds the illusion of going up to roost. Others are set on poles above ground level and have to be accessed by a ramp. An attached run is more useful if it is at least partially covered by fine mesh tarpaulin, to protect the birds from adverse weather conditions.

The availability of a designated run can be useful in the day-to-day management of the flock. If birds can only enjoy the freedom of the garden when a member of the household is on hand to keep an eye on them, and that time is limited, they will also need access to a secure run, even if that space is only small, when the householder is not present.

Free-standing, static and free-range houses

Large houses that hold large numbers of birds, for a small-scale egg business perhaps, include those where the occupants range within an enclosed run that is integral to the structure of the house, or in which a static hen house is placed in a designated area with a larger surrounding area enclosed by fencing. The area between the two provides grass in which the occupants are free to range. In a totally free-range system, it is

▲ *Eggs removed from a large house by a conveyor belt system are kept cleaner and are easier to collect.*

▲ *For those with a small field, a house with a slatted floor may be a good choice. If the housing is moved regularly it will halve the cleaning time. Regularly deposited droppings can help maintain overall soil fertility.*

▼ *These large fowl, kept in a static house and run, will soon make the solid base dirty enough for it to need sweeping clean. Miniature pure breed fowl could also be accommodated here.*

possible to see poultry of varying ages and sexes apparently co-existing happily. Each individual, or sub-group if the flock is large, will find its own space.

Many purpose-built static systems are ideal for hybrids as well as the more productive bantams. The land on which the poultry range is not necessarily of natural materials. This category of housing also includes permanent or semi-permanent

structures containing intensively housed poultry that are rarely allowed to venture beyond its confines. Some sophisticated sheds used to house laying hens have nest boxes that allow the newly laid eggs to roll gently into a collection tray.

Static houses may be the only solution in an urban location where the owner wants to house a large number of birds in a designated space. A lean-to housing design that was once popular among post-war domestic poultry keepers can be adapted to fit a range of situations.

Permanent runs

The base of any permanent run should be slightly higher that the land around it so that water does not flood

▼ *Tiny, true bantams that are kept as pets will, if given lots of attention, live happily in relatively small pens.*

in from the surrounding ground. A solid or concrete surface will require a shallow covering of coir, hemp or even fine wood shavings, and these will need changing on a regular basis. A free-draining base, raised 30cm/12in above ground level, can be topped with 30cm/12in of untreated shredded wood or bark. This does not need regular changing, only an occasional rake over. With much attention from the hens, this lining material will gradually turn into a valuable compost product that can be replaced as it either begins to break down or starts to smell. Poultry runs will benefit from being wholly or partially covered to provide shelter from the weather.

All of the static run systems require more cleaning than movable poultry units. Owners with extensive runs, as required for medium-sized, free-range egg-producing ventures, will have to construct an efficient fence system capable of foiling the most persistent of hungry foxes.

▼ These specialist table fowl enjoy searching through this over-long grass for grubs and insects.

Secondary housing

Most poultry keepers are, at some time, likely to require some form of secondary housing in which to isolate ailing or bullied hens. Even owners keeping three or four hens should make some provision for such birds. Nearly all the purpose-built laying houses are built for healthy hens, and have no provision to screen off some of the area for hens that need isolated respite care. A second small-scale,

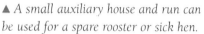

▲ A small auxiliary house and run can be used for a spare rooster or sick hen.

general-purpose housing unit, with a free-range run that includes a nest box, will be essential for people keeping larger numbers of poultry. Modern plastic housing, which is hygienic and small-scale, would provide an excellent example of a general purpose or isolation unit.

On a simple scale, a poultry sickroom could be a very large box with a mesh top for access or the

▼ This house has a big enough door to accommodate a large exhibition-standard Sussex, but may not have sufficient ventilation on a hot summer night.

provision of a heat lamp. However, a wire-gated front would give far better human access.

Over the years, the "broody house", usually a small A-framed house, has found a range of uses beyond its original role as a sanctuary for the broody hen and her chicks. The best examples of these had a sliding panel in the roof and removable bars at the front enabling eggs to be carefully placed under a sitting hen, or to give the sort of access needed to tend a sick fowl. In a modified form, when raised off the ground and given a slatted floor, a similar little house could be used to "break" or return a broody hen to laying mode.

Roosters do not need to be kept continually in the breeding pen, so a larger house with a run could be used to give him or his hens a rest from too much sexual activity. A small ark can be used as additional housing to hold a troublesome growing rooster for a few months in the summer until a new owner can be found. It could

▼ *Small-scale housing, more usually associated with rabbits and guinea pigs, is also suitable for bantams.*

also be used to give one or two exhibition bantam hens or a bird recovering from a health problem access to the best possible show conditioner and tonic – grass.

Housing exhibition birds

The basic principles of housing breeding fowl, or large exhibition fowl, remain the same for hybrids, bantams and standard pure breeds. Protection from the rain and puddles,

▲ *Providing plenty of light, a layer of clean straw and a droppings pit under perches covered by strong mesh means that these utility-type Sussex can be kept inside during a snowy spell.*

▼ *Many exhibitors use the same top-quality, white wood shavings that are used in exhibition pens on the floor of the poultry house. Others use either shredded hemp, or occasionally, wood or paper pellets.*

▲ *Part of their regular routine may be a period spent on grass, but each of these Belgian bantams will wait patiently to be picked up and taken to their run.*

▲ *This wire front of the house (removed) allows good ventilation. Good husbandry and lots of attention means that these bantams are content in a confined area.*

good ventilation and security from predators are paramount. Breeds such as Orpington and Sussex are now far bigger and fluffier than at any

▼ *Having been handled since they were one day old, many show birds look for daily contact with their owner*

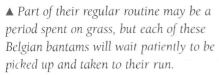

previous time. Such large exhibition varieties have become so big that their size limits the options for suitable housing. Instead of modifying large poultry houses, breeders of such fowl have adapted garden sheds and even stables for poultry housing. Some exhibitors, particularly of bantams,

keep their fowl in an enclosed environment for regular inspection and handling. Such breeders may move their birds through a succession of indoor hutches and pens to keep their fowl healthy as they grow. Others will introduce their young stock as early as possible to an outdoor environment.

▼ *These bantams may spend time in a grassed run, but still see the surroundings as an alien environment.*

BUILDING YOUR OWN CHICKEN HOUSE

There are many reasons for building your own poultry house: to cut the cost of a pre-packaged product; to use good DIY skills alongside easy access to timber; perhaps for the need for a specific design feature within the house; or to accommodate poultry with special requirements.

Before the advent of quality ready-made housing, almost all domestic and small-scale poultry keepers would have made or built their own. While it is unlikely that you will make a house as cheaply as the price of a lightweight flat-pack, there are still instances where making your own may be the best option.

There are gaps in the market for specific types of housing. If the type of housing that you require is not available, then building your own may be the only option. Many features once incorporated into the designs of poultry housing are no longer included today, or modern concepts of how poultry should be kept means that they are available in

▼ *The poultry house made from recycled timber will last as long as a shop-bought model if it is well looked after.*

less than satisfactory versions. Small broody houses, soundproof rooster pens, and larger housing, tall enough to walk through while still being able to accommodate some of the perching occupants, are a few of the less easily obtainable options.

Additionally, people considering keeping a few hens but who are unsure if poultry keeping is really for them could think in terms of converting a budget garden shed into poultry housing. This conversion should be totally compatible with hen welfare, and if you decide the hobby is not for you, it will still be possible to return the structure to normal garden use. A checklist of the pros and cons, including the costs and type of material used in converting or building your own poultry house, could help you decide whether to buy ready-made housing or build your own.

Thrifty use of materials

The post-war period saw an enormous expansion in domestic poultry keeping. With resources scarce, the majority of poultry houses were homemade, and were nearly always from recycled timber. For those interested in the conservation of a scarce resource, using second-hand materials to make a poultry house is still a viable option.

Parts of the house, or even a small house or broody unit, could be made from new lightweight timber off-cuts if you have access to them. A visit to a reclamation yard may yield opportunistic purchases which may require modification to make them suitable for the planned build.

Bear in mind that many fibreboards are not weatherproof, and all chipboards provide ideal sites for mites, so choosing materials for side cladding and the roof needs careful consideration. It may be possible to treat crevices between boards with mite powder, but not when they are covered by roofing felt; so, while plywood sheets may make an excellent roof, any joints should be treated before being covered with waterproof material.

Aside from cost, the obvious advantage of making your own poultry house is the ability to tailor it to make the best use of the space available. In addition, you will be able to adapt the internal features according to the type of poultry that you wish to house. Many exhibition, large, or fancy poultry breeds have specific requirements not catered for by standard off-the-shelf poultry housing.

Practical considerations

The steps that follow guide you through the stages of making a poultry house or adapting an existing structure, such as a garden shed, so that it is suitable to house poultry. The house shown is a permanent structure, with a trapdoor to allow the poultry access to an enclosed outside run, and also a door to allow human access. An opening is cut into the front external wall of the poultry house, and a nesting box is fitted to the external wall to cover the opening. The nest box has a lid to allow access to collect the eggs from outside.

The house front has a wire mesh-covered "window" to allow light and air into the house. At night, the top of the mesh window can be covered with a drop-down door front, which is held on a hinge. The bottom of the mesh window is covered with removable wooden panels that are held in place with clip over fasteners.

Internally, the house has a solid floor, a bench, staggered perches on which the poultry roost at night, and easy access for the occupants to the partitioned nesting box area.

This poultry house has been made from reclaimed timber and offcuts salvaged from other projects. It has been clad in shiplap panels, and requires good DIY skills to make a similar structure, or to adapt the plans to suit an existing wooden building. The basic design here assumes that you have or can make a wooden building with four walls, a floor and roof. The building shown has been assembled in a workshop, then dismantled and reassembled in situ.

Choose your final site, ensuring that the ground is level, digging out and filling with scalpings (rock waste) if necessary. The house could stand on treated timber bearers running the length of the building to help preserve the floor joints.

1 This view shows the front exterior wall of the poultry house, with the nesting box area in place, complete with a lid to allow the owner access to collect the eggs and clean out the nesting area. A hole has been cut into the lower right-hand side of the exterior wall near to the nest box area, to allow the poultry access to the house. Above the nesting box is the "window opening", which will be covered with wire mesh and possess removable doors.

2 An internal view of the back wall of the poultry house. The back wall has a bench on which the hens can sit, and staggered perches set above bench height on which the poultry will roost at night. The bench acts as a droppings board. The flooring material is nailed to wooden beams, allowing air to circulate beneath, and a doorway is visible on one side wall.

To make the nest box

Nest boxes can be tailored to suit the occupants of the hen house. Hybrids and most bantams may fit into a 23 x 25cm/9 x 10in space, whereas a large Indian Game bird may need double this space. The four nest boxes added to this poultry house are made of a single length of timber offcut, which is partitioned to make three internal walls. A timber lid, floor, two end walls, and battens to attach the nest box to the exterior house wall are also required. Plastic nest boxes can be obtained for inside the hen house.

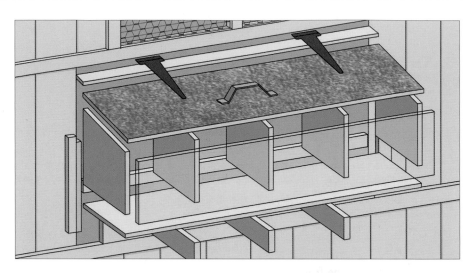

1 Nest boxes have been created by adding three internal walls to the nest area. Battens strengthen the house wall and hold the nest box in place.

2 The nest box area extends from the exterior wall of the house. The lid allows access to collect the eggs without having to enter the house.

3 The lid is attached to the exterior wall of the house and to the partition walls with a length of timber nailed or screwed in place.

4 The rough edges of all timber boards, should be sanded down to avoid owners or birds being harmed by splinters and to make cleaning easier.

5 Nest boxes can be a haven for red mite, and chipboard can provide crevices for them and their eggs, so choose materials carefully.

To make the dropping board and perches

This house has been designed so that the poultry keeper can walk inside and handle the poultry. Such easy access allows the owner to monitor the birds' living conditions and allows him or her to adjust the ventilation as necessary to maintain an optimum living environment for the poultry. The droppings board or bench is set below the perching area. The perches need to be tailored to the size of occupant; the bird grips the perch with the central toe, which is longer than the others. For greater comfort, the front and back edge of the perch should be slightly rounded.

1 The droppings board sits on top of battens which are fastened to the internal walls of the hen house. This helps to keep the floor clear of debris.

2 The board is cut from a single piece of timber. The fewer pieces that make up the house, the fewer places there are for mites to live.

3 The board is fastened to one wall, and is removable to allow deep-cleaning of the house. Line the board with old newspapers, plastic or cardboard for easier regular cleaning.

4 Some large breeds of poultry can be heavy, particularly if a few birds stand on the board at the same time. The space beneath can be used for feeding stations in extreme weather.

5 The perching area is made as a separate unit and is attached to battens fixed to the internal wall. The angle and spacing should allow for the optimum number of birds to perch.

◄ 6 At each end the perches attach to a batten. Right-angle cuts have been made into the batten face to accommodate the perches. Perch ends are a haven for red mite.

► 7 Fit the diagonal battens to the battens on the end walls and hold them in place temporarily with a nail to check the fit. Ensure that the perches are level and have a smooth slightly curved surface for the poultry.

Additional internal features

The roof is a flat surface that fits inside the top of the exterior walls like a box lid. Short battens attached at equal intervals around the internal house walls will support the weight of the roof.

The front of the poultry house has a window for ventilation, a feature that is significant for the wellbeing of the poultry and which is often inadequate in many ready-made or flat-pack poultry houses. The window is securely covered with chicken wire, to keep the birds in and predators out. The wire edges are covered with a thin batten.

1 Roof supports are fixed to the internal walls. Shelf brackets provide additional reinforcement. The roof joists will sit on the supports.

2 Fasten the wire mesh over the window aperture and secure in place with battens, to ensure that sharp edges of wire are not protruding.

External features

The top half of the chicken-wire window is covered by an adjustable shutter, which is attached to the exterior wall of the house with a batten and some hinges. Use a traditional window fitting to prop the window open in daylight and to allow maximum airflow through the house. Two smaller detached doors with handles cover the lower half of the window. These are held in place over the window with clip fastenings at night and in poor weather conditions. Sliding glass windows could also be used. Alternatively, breathable scrim provides ventilation in warm weather.

1 To complete the exterior, cover the nesting box area with roofing felt and add a handle.

2 The adjustable shutter provides additional ventilation in summer and when closed, keeps the house warm.

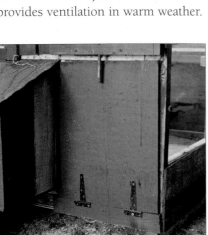

◀ 3 The trap door is made from a timber off-cut and is attached to a batten, which is secured to the lower edge of the house with hinges. A bolt ensures that the door can be locked to keep the poultry secure at night. This drop-down door acts as a ramp for the poultry to enter and exit the house.

▶ 4 Finally, paint the exterior of the house with exterior timber paint.

Disassembling and reassembly

For practical purposes and during wet weather, the poultry house is probably best made and assembled in a large covered area. Each section can be temporarily held in place with nails hammered half-way into battens, to ensure that all essential components are present and that the house works at a practical level. The poultry house could also be assembled in situ, depending on the weather; you will need to make sure you have all the components ready. The panels are permanently secured using screws, then the house is clad in weather boarding, treated to prevent mites.

1 Once the exterior wall paint is dry, remove the internal features and each wall panel to disassemble the shed.

2 In situ, the house stands on railway sleepers to keep the floor dry in wet weather.

3 Ensure the floor joists are level. Once the foundations are adequate, the structure will stand securely.

4 The house stands on beams. The underfloor cavity should be blocked off to keep the poultry out.

5 Put the floor in place. The underfloor cavity is an ideal place to position a plastic tray of rat poison.

6 Two people will be needed to lift the component parts of the poultry house into place while they are attached.

7 Check all walls are straight using a level, and that they are secure. The house also needs to be watertight.

8 Cover the roof with roofing felt, nailing it in place according to the manufacturer's instructions.

ADDITIONAL ESSENTIALS

Nesting boxes, perches, feeders and dust baths are essential features of the hen house that can be made or purchased to maintain the wellbeing of the poultry. Lighting will help to secure a regular supply of eggs, particularly during the short days of winter.

Poultry houses must be fitted with nest boxes in which the hens can lay eggs, as well as perches on which the birds can sleep in a darkened area, and feeders in which food, grit and water are supplied. There should also be as much natural light as possible inside runs and feeding areas.

Nest boxes

A nest box is essential for egg-layers. Usually one box shared between three or four hybrid hens in full lay is adequate, but many traditional breeds of large fowl will require larger dimensions than the standard 30 x 35cm/12 x 14in allowed in ready-made houses. The nest box should be filled with sawdust or hay to encourage nesting, and should ideally offer human access from the outside for egg collection. Some provision to stop hens sleeping in them should be built into all nest boxes. A properly constructed nest box will have a partition that denies pullets access while they are being trained to sleep

▼ *Other than in the most secure runs, most poultry keepers will have to close the trapdoor at dusk and open it again in the morning.*

on the perches, otherwise they may get into the undesirable habit of sleeping in the nest boxes.

Perches

All poultry require a sleeping area, which should be a darkened area that does not allow in natural light. Birds

▼ *Birds will soon learn to mimic their tree-living ancestors and climb a ladder in order to roost at night in their sleeping quarters.*

▲ *Feeders should be held at roughly crop height and should be adjusted to allow a feed level of about 1cm/½in.*

obey a natural instinct to settle in for sleep as soon as it starts to become dark outside. If they are not locked into the house early enough, many birds may fly up to a suitable perch

▼ *Birds unable to create a natural dry dust bath will need a bowl or tray of suitable material to dust themselves in.*

▶ *Birds may enjoy a natural bough as a perch in an outside run, but will require one of appropriate width to spend the night on.*

outside the house, which may make finding them difficult, as well as making them vulnerable to animal predators.

For sleeping, all birds require perches, which should be positioned appropriately for the size of bird you wish to house, and should always be above floor level. In most hen houses, the perches are ideally suited for housing hybrids. However, tiny bantams will find the perches provided to be too wide, and a giant Shamo, too narrow. While 25cm/10in of linear perch space may be enough for hybrids, twice this length would not be adequate for a large, feather-footed Brahma. Most fowl may be able to jump or fly 1–1.25m/3–4ft to roost, but an Orpington or Indian Game may need a lower perch.

▼ *Nest boxes should have a fold-down partition to help train pullets not to sleep in them.*

Feeders

The provision of some sort of trough or feeder is essential inside the hen house. As an increasing proportion of laying and growing poultry are now fed "on tap", many people choose a tubular feeder that can be adjusted so that a small amount of feed is always available, while at the same time, spills are kept to an absolute minimum. Wherever possible, feeders should be suspended to about crop height and preferably sited within a covered part of the enclosure.

A constant supply of fresh water is as important as a source of food. Very large poultry units may have an automated supply, but most small-scale poultry keepers will rely on galvanized or plastic founts. These are readily available at most feed barns or pet shops, but most of them rely on the water being delivered via an opening that never seems quite large enough. More experienced users ream this hole to be one or two sizes larger, in order to stop it from being blocked by small bits of chaff. An 8cm/3¼in hole allows free passage of water. Use a translucent container so that you can see when to refill it.

Dust baths

Hens that are free range always seem to find an area dry enough to create their own dust bath, even in wet weather and in the most unlikely soil conditions. Taking a dust bath is the

▼ *A hygienic plastic nest box can be fitted with a rounded base insert that allows the egg to roll to the front for easy collection.*

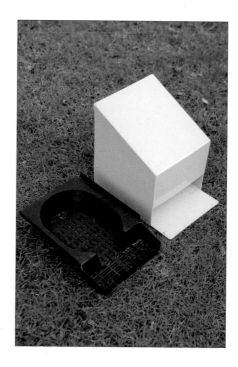

method by which fowl rid themselves of body parasites. Even when they are completely vermin-free, the daily dust bath seems to be an indispensable part of their routine. Hens housed in small runs should have access to a box or bowl of wood ash, fine garden soil or even Fullers Earth.

Interestingly, the very fine dust created by the hen in the dust bath may not be dissimilar to that now used as crevice treatment against red mite.

Lighting (optional)

In the 1880s poultry keepers first noticed that hens kept under the glow of urban gas lights laid greater numbers of eggs than their country cousins. Since then, humans have experimented with providing extra lighting for laying fowl in the hope that they will be tricked into believing it is daylight and lay more eggs.

Today, most industrially reared pullets live under a controlled light regime in order to be most productive. Similarly, both battery- and barnyard-laying hens are usually kept under a regime designed to

▼ *It is control of lighting that accounts for much of the commercial success of hybrid egg-laying poultry.*

obtain maximum output of eggs throughout the year. Traditional breeds, as well as modern hybrids, that live according to natural daylight hours are expected to lay less frequently during the winter, when there are fewer daylight hours. This is because the poultry have difficulty in consuming enough food in daylight hours to sustain themselves properly during up to 17 hours of darkness. For this reason, many small poultry keepers will consider providing electricity to the poultry house, in order to light the house during the dark winter days. A modern time-switch can be set to allow an extra two hours of light per day, which will enable the birds to feed, preen and socialize. This improves the birds' welfare, and also increases the number of winter eggs laid. However, if you keep a rooster and add these extra hours in the morning, you risk receiving complaints from your neighbours about the crowing.

HYGIENE

Practising good hygiene in the hen house and in a static run will keep the environmental impact of keeping poultry to a minimum. All poultry keepers will want to limit smells, flies, noise and particularly vermin, but make sure any rat poison is placed out of reach of other animals.

Keeping too many birds in an uncovered wet run can soon create a foul-smelling, muddy mixture of soil, faeces and spilt feed. This may increase summer fly activity, and attract the interest of visiting rats. Moving the poultry run regularly to allow the hens access to fresh grass can help to limit problems and increase the poultry's wellbeing. Grass that has become soiled or been subject to heavy use by a few hens in a confined area can look unsightly, but will recover. The normal action of grass re-growth and earth-worm activity will quickly disperse a few days' worth of poultry droppings into the soil. Droppings from a static poultry house can be placed on the compost heap and will increase the efficiency of the composting.

If waste feed is kept to an absolute minimum, poultry may attract no more rats than the average garden bird table or compost heap. A precautionary rat-baiting station can pre-empt an unwelcome infestation.

▲ *If not properly cleaned and maintained, a static run can become smelly.*

Pest-free housing

If the air flow in the poultry house is correct and the birds are healthy then any odour will not be that unpleasant. Providing the occupants are free from dietary upsets, even a build up of manure in a droppings pit need not be too noxious. Dust-free cobwebs in the house can indicate air movement rather than draughts. If the birds have clean nostrils and unlaboured breathing the ventilation is appropriate.

▲ *Muddy conditions can be unpleasant for both hens and owners.*

Crevices that could harbour mites need to be kept to an absolute minimum. Having said that, the more accessible the crevices are, the easier it is to coat them with one of the safe, modern, contact preparations. Regular visits to a hen house should mean that you are the first to be aware of any flea, mite or louse infestation. React quickly if there are any problems, since these pests can spread quickly through a flock.

▼ *A tell-tale new rat hole is an early sign of a potential infestation, and will need to be dealt with swiftly.*

▼ *Regular precautionary baiting against rats in a safe location is important. An anti-coagulant poison is a favoured option.*

▼ *Mink are vicious predators that will prey upon poultry, and create havoc in a small backyard.*

HOUSING YOUNG POULTRY

Breeders who rear chicks have a specific set of housing requirements for their stock. Other than those brooded naturally under a mother or foster hen, chicks will need some form of additional heat. Even those brought up by their mother will require protection from predators.

Aside from poultry keepers who hatch and rear chicks under broody hens, most breeders will start rearing their chicks under some form of heat in a cosy shed or outbuilding. This will replace heat which would otherwise be provided by the mother hen.

A cosy coop

As the temperature outside the chick enclosure may rise or fall significantly within a few days or overnight, most chicks require significant insulation. Most brooder rooms are less than perfectly insulated, so it makes sense to erect an insulated igloo around the birds to enclose them, especially if heaters are being used in a given area. Some of the heavier commercial strains grow so quickly that, while requiring supplementary heating, they

▼ *The little apex ark can make a perfect first home for a mother bantam hen and her chicks.*

may also need re-housing. Strains that feather early may be ready to move to an unheated house at three weeks old in the summer, while slow-feathering examples may require an insulated winter home for two months or more.

The difficult transition for the chick comes when it reaches the stage of a just-feathered grower ready to move outdoors. This usually occurs any time from spring to autumn, and requires specialized or adaptable housing if the young are to survive inclement temperatures.

Growers to point of lay

By far the best and most universally popular rearing system used is based on the A-framed ark design. The original, simple Sussex slatted-floor ark used to be the house of choice on most small poultry farms. The slatted floors allow droppings to fall to the ground and stale air to pass upwards

▲ *These newly hatched chicks are housed in an outbuilding, and are retained within a specific area by a hardboard enclosure, which helps to keep the heat in and draughts out.*

and out of the roof vent. With a layer of straw on the floor and the corners insulated with hay contained behind stiff cardboard, this would make a snug home for 40 one-month old chicks. During cold periods, the whole house can be insulated by putting cardboard over the slatted floor. On really cold nights, the young birds could be made cosier by installing a false roof made of wire netting that supports a thick duvet of straw or dust-free hay, or by hanging sacking over the wire pen front.

Stripped of this insulation, the slatted floor, apex roof and sliding wire door of the A-frame ark ensure perfect ventilation when outdoor temperatures rise. On sunny days the solid roof door, which provides human access to see inside the house,

▲ *As growing fowl feather, they may require little extra heat but will quickly outgrow their living space.*

can be replaced with one clad with clear corrugated PVC. The same material can be used to cover any small run used to both shelter and contain its younger occupants.
As the birds grow, the ark can house 25 pullets until point of lay.

▼ *Early-feathering chicks are off heat but still kept warm by a false roof and blanket of straw.*

▲ *Growing pullets explore their new straw-littered home and begin to forage as in nature.*

Moved a short distance every day to fresh grass, the droppings fall on the ground to be absorbed back into the soil. Birds reared in this sort of house feather better and easier than in most other types of housing.

▼ *A circle of hardboard within the chicken house restricts the poultry to a small area. Straw is packed in behind to help create warmth for the birds.*

HYGIENE

Chick droppings, that previously dried quickly under the dry atmosphere of a heat lamp can suddenly increase in volume and stickiness as the birds grow and produce greater volumes of waste. In summer, if the waste is left to build up, it may create the sort of warm, damp heat environment in which coccidiosis spores multiply – with potentially disastrous results.

In winter, lower temperatures can soon cause the whole brooding area to become damp and smelly.

Daily cleaning of the housing area can be made easier by covering the floor with sheets of corrugated cardboard. This has the advantage of increasing the insulation in the area, and can simply be rolled up and burned when soiled. The same technique can be employed when growing birds are moved to their more permanent or outdoor quarters.

▼ *Insulation provided by straw will provide slow-feathering breeds with warmth in cold weather.*

SECURITY

With more poultry being lost to foxes than to any other cause, careful consideration needs to be given to ensure the security of the inhabitants of the hen house. The same precautions will help to deter other potential predators, such as cats, dogs, mink, badgers, and even human thieves.

Threats to poultry include wild mammals such as badgers, mink and members of the weasel family, feral or pet cats and dogs, and of course, the wily fox. Especially when driven by hunger or the desire to feed cubs, foxes are intelligent, resourceful and agile. They are quick learners and natural opportunists that will soon take advantage of slack security on the part of the poultry owner, or a temporary lapse in caution. In the USA, poultry are also at risk from coyotes and raccoons. Foxes quickly establish where hens are present, so even though you may not have seen a trace of a fox, this nocturnal animal is guaranteed to know where your stock is kept. Foxes are clever and adaptable; leave the hen house open only once, and the chances are that by morning most of the occupants will have been slaughtered. Foxes don't take just one hen for food. Even if it takes one bird to feed its family, the fox will not leave the house

without slaughtering the other birds. And sooner or later, most unfenced hen runs will be subject to a daylight raid, often by a hungry vixen or her growing cubs. If hens are allowed to roam free in the garden, they need constant supervision in order to deter the fox, since the urban fox has lost its wariness of humans.

Apart from the need to keep out the fox, as well as human thieves, free-ranging hens need to be kept from straying too far from home. Rounding up hens that have made a break for freedom might be a novelty at first, but it quickly becomes tiresome, and can be annoying to neighbours, especially if the hens pull up prize plants. Aside from this, the greater the ranging area, the further you may have to look for the eggs. A downside of hens that are free to roam is that they take every opportunity to hide away eggs that may become stained and unsaleable. The answer is to construct a fence

▲ *The increase in interest in pure-bred poultry has been matched by an increase in thefts from exhibitions and breeder's premises. Improving security may deter opportunistic theft.*

around a free-standing house to keep the hens in and the fox out.

Given an hour or so, a fox can quite quickly dig its way under a wire fence, but will usually be defeated by netting that is folded outward at ground level for a few inches before being buried the same distance in the ground.

Outfoxing the predator

Substantial and extensive fence systems are an expensive and long-term investment, so many poultry keepers may look at the alternatives

◄ *Foxes are nearly as agile as cats. They will find it easy to use a taut wire fence as a ladder and climb into the house, whereas a less rigid fence will deter them.*

SHORTENING POULTRY WINGS

Some of the lighter breeds of fowl are capable of flying over fencing, and may need to have the primary feathers on one wing painlessly shortened, a procedure that leaves the birds sufficiently out of balance to deter them from flying. The procedure may have to be repeated after the annual moult, and can be carried out by anyone competent in handling poultry. Use a pair of sharp, clean scissors to clip the primary flight feathers to half their original length.

▼ *Before shortening.*

▼ *After shortening.*

before being convinced that such an outlay is necessary. Electric fence systems can be used to stop predators from burrowing under or scaling a permanent fence. In a modified form, they can be used to keep the predator at a distance from the small run. All electric fence systems rely on a fox or coyote (in the US) having once been given a sharp shock, thereafter treating any similar wires with some deference. The message to the fox can be reinforced by applying some strong-smelling liquid. Cheap scent, timber preservative and human urine have all been tried. The same product can be used on a temporary house to keep a fox away from an isolated sitting hen.

Plastic and wire mesh fence systems can be used effectively to deter predators, even though the bottom of the fence and some of the vertical wires are not electrically charged. There are few reports of

▶ *An electric fence along a perimeter fence may soon deter all the local foxes.*

animals jumping over or burrowing under this sort of fence. However, some hens will either fly over, or worse, get a shock on their way out, and then be reluctant to make the return trip. Some larger poultry units invest in a dual system of traditional perimeter fence reinforced by a single low strand of electrified wire. The

way that foxes approach a permanent fence means that an electrified wire positioned 15cm/6in from the fence will deter them from attempting to climb over, or burrow under, the obstruction.

All electric fences rely on an energizer to either transform a small, low-voltage battery supply into a sharp, stinging shock, or mains voltage scaled down to a similar but still harmless level. The efficiency of the system and the distance over which it will be effective is determined by how well the energizer's second wire can be earthed. Even a single electrified wire, if kept in situ can act as a deterrent to a fox in the same way that a similar but smaller output unit, powered by a battery, can be used, to energize a bungee wire system. Some of the more powerful fence energizers may scorch young grass, so any vegetation in close proximity to the wire will need to be regularly trimmed or sprayed.

▼ *Predators make unwelcome visits to poultry houses, but will soon learn to associate a chicken house with an unpleasant experience if a small-voltage electric shock is received.*

ROUTINE CLEANING AND MAINTENANCE

Hygiene and ease of cleaning are obviously priorities when choosing poultry housing. Regular cleaning and maintenance will not only ensure proper hygiene but also prolong the life of housing and runs, making life more pleasant for the occupants.

Poultry and their living environment must be kept as clean as possible so that the eggs they have laid arrive in the kitchen or hatchery in prime condition. In any case, if poultry live close to humans, an unsavoury environment that smells unpleasant will affect their owners too.

The easier a house is to clean, the more often it is likely to be cleaned, so this is important to consider when purchasing a house. A housing system in which the poultry spend a lot of time in a run or a grassed enclosure may take less time to clean than a static house in which the birds spend most of their time indoors. A house that has an integral grass-covered run and can be moved regularly to a new site may be the perfect choice, that is, until wet weather creates quantities of mud that is carried inside by the occupants.

Cleaning the house

Modest numbers of growing chicks in a house soon produce enough damp faeces for their bedding to need changing on a daily basis. Housing with slatted flooring that allows most of the debris to fall between the slats has much to recommend it, but only if the slats are cleaned on a regular basis.

Increased quantities of soya protein in poultry feed is often blamed for poultry droppings becoming sticky and gooey, making the chore of cleaning harder. Domestic pressure washers are useful cleaning tools. If all the poultry are disease-free, then regular household cleaning products will suffice for cleansing poultry housing. However, when cleaning to

▲ *Pressure-washing slatted floors is a routine task between housing batches of growing poultry. Disinfectant may be applied at the point when all organic matter has been removed.*

contain an outbreak of disease, a specialized germicidal preparation should be used.

Many cleaning tasks can be simplified, for instance, cover areas under perches with either small sheets of polythene that can be hosed off, or corrugated cardboard that can be composted. Alternatively, certain areas can be covered with an inert material: wood shavings, absorbent wood pellets, or more readily compostable straw and hemp products.

Keeping eggs clean

Dirt can help germs to penetrate eggshells, so keeping eggs clean is a priority, whether they are for eating or

▲ *This Sussex ark, the universal rearing choice, will give years of useful service if regularly cleaned and preserved.*

hatching; collect eggs regularly. Chicks should ideally hatch germ-free. However, washing dirty eggs has been shown to help germs to penetrate the shell. Approved germicidal products are available to sanitize eggs intended for incubation. As hens rarely foul the nest box, a handful of inert material that is changed on an almost daily basis can go a long way to keeping it and the newly laid eggs perfectly clean.

Maintenance

An annual or bi-annual application of a modern wood treatment will help to prevent rot in woodwork that has not been pressure-treated, but will not control red mite, which live in crevices in the house.

▲ *Spraying crevices to kill red mite must be done on a regular basis to keep the pest at bay.*

▲ *A strong wire grille enables poultry to walk over the top of a droppings pit under the rack of perches.*

▲ *One of the new plastic nest boxes complete with pull-out liner can ease the work of keeping the eggs and nest clean.*

If you discover an infestation of red mite, check to see if they have spread, and re-roof the house if necessary with other materials.

Good husbandry can extend to cultivating and replanting runs with fresh grass if it has become stale or unsightly through overuse, or to provide a crop of fast-growing foliage that may be used to supplement food.

Most small-scale poultry keepers who keep their birds within a confined and shared area will opt for ultimate cleanliness. This means that floors are kept clean and covered with wood products which are inert and slow to break down. Other owners with larger gardens may view poultry droppings as the basis of an active compost system which will ultimately

be used to grow food for the poultry and for their own consumption. Here, a regular cleaning routine will be just as important as in any other system. It is possible to line the floor of commonly used areas under perches with a sheet of cardboard. This can be removed to the compost heap, where it will quickly break down into usable compost.

Changing the bedding

1 Straw or softer hay is the traditional poultry litter used to line the nest boxes, but many owners now prefer wood shavings.

2 Used straw and the accumulated droppings are a valuable part of the garden compost heap. Bales of straw can be bought from suppliers.

3 Rather than using a large quantity of straw or hay, it is better to use a small amount that can be changed as soon as it becomes soiled.

CARING FOR YOUR POULTRY

Good husbandry is imperative to every successful poultry enterprise, whether you have one or two birds in the garden or a small flock from which you sell eggs. Keeping the fowl healthy and their living conditions in good order makes for content hens. Observing and understanding the young bird's natural behaviour, and providing them with an environment that promotes that behaviour, is key to their wellbeing. Hens in good health need less intensive care and attention. They can also add significantly to the garden economy, as well as promoting your own pleasure at spending time in the garden. Not only do hens provide you with eggs, if left to range free they can play a role in the garden's ecosystem. Allowing poultry to eat any crop residue from the vegetable garden increases their dietary range. They, in turn, will provide manure beneficial to the health of the soil. Of all the attributes that make a good poultry keeper, the most significant is the ability to recognize potential problems and nip them in the bud. This means carrying out regular routine inspections of each hen. Learning how to handle them correctly and knowing what to look for will make this task easier.

▲ *A large poultry feeder will allow numerous birds to feed at the same time and should counter any anti-social behaviour.*

◄ *Dark Dorkings enjoy a free-range and natural environment, one that is ideal for most breeds of poultry.*

POULTRY BEHAVIOUR

If you keep a few hens in the back garden you should aim to provide an environment that promotes their natural behaviour. It is essential for a hen's welfare that it is allowed to live according to its instincts, and this is key to any organic or free-range environment.

Should you leave a hen to its own devices in a free-ranging environment, it will spend its time grazing, pecking the ground, scratching in the earth and bathing in the dust. However, unless you are around most of the time to keep predators at bay, most poultry will spend a significant amount of time in a run, and be let out only occasionally to roam in the garden.

Poultry are social birds with a strong social instinct to interact with each other. They naturally live in a small flock, and require the company of other birds, no matter how few. Birds naturally roost together and forage separately. A flock creates the ideal environment to provide each hen with a sense of safety. Within the flock, a pecking order will be

▶ Some hens habitually choose awkward places to lay their eggs.

established in which stronger, dominant birds top a hierarchy, and birds with a more passive nature may be bullied. However, birds will not necessarily bully each other unless other factors promote it, such as inadequate space at feeding stations or the introduction of new and unfamiliar birds to the flock.

Ruling the roost

More than one male may be able to co-exist with females in a flock without roosters fighting each other for supremacy, but much depends upon the breed. Males that have been brought up together are more likely to live companionably, but some strains of game breeds are capable of killing even adolescent siblings. Senior males that have a perceived entitlement to remain at the heart of the flock may banish younger roosters

to the periphery. If a batch of young males in a flock starts to become aggressive and establish a pecking order, an older rooster put with them will act as a "policeman". If there are enough hens to go around, all should

▼ Hens peck for tiny insects and even yeasts to augment their diet. This habit also prevents boredom.

▼ He may be small, but being older, this bantam rooster uses this seniority to keep the peace among the much larger adolescent growers.

live companionably. Since many egg-laying flocks are all female, older female hens may also rule the roost in the same way as do aggressive males. A hen or rooster that becomes too dominant can be separated from the flock by being kept in different housing that is still within view of the flock. A trial reintroduction after a few days will establish whether the bullying behaviour has been circumvented.

Feather pecking

This problem is at its worst during the long daylight hours of summer, when poultry become bored. The dominant birds may turn their attention to a more docile bird that is lower in seniority and peck at it, removing patches of its feathers. A flock of birds may also "attack" newly introduced birds, or peck an unhealthy bird, even occasionally to death. Such aggression may be upsetting to witness and seem barbaric, particularly if you imagined happy birds clucking around your garden. This is a behaviour pattern

that affects egg-laying poultry, and is a serious problem for the poultry industry, particularly where birds are intensively housed.

It is thought that feather pecking occurs because the food that the birds are provided with is either so complete that it frustrates their natural inclination to peck for food, or that it is deficient in minerals and nutrients, resulting in antisocial behaviour. Feather pecking has been seen to lessen if the birds' diet is changed to a high-fibre one that the birds have to keep eating for longer in order to get the nutritional levels that they need, leaving less time for boredom.

Groups of differently coloured hybrids or crossbred layers housed in a small run may also experience problems. The narrow run means that the back of each fowl is always in front of another bird, particularly so if the birds are semi-intensively housed. While the birds may have been reared together and initially cohabited comfortably, on moulting they see that each bird has differing colour feather quills, and inquisitive feather

▲ *A cabbage hung at head height or higher can supply valuable nutrients, and help prevent boredom, bullying and feather pecking.*

picking may begin, which soon progresses to pecking. During the short winter days, there is also competition for trough space, which may create additional problems.

For most of the year, dry mash in the hopper occupies the birds for longer than pellets, thereby reducing the time for them to get into mischief. In a confined space, a cabbage hung at head height rather than thrown on the ground can be of interest to the birds, but may leave some looking for something else to peck.

The problem can be partially resolved by reviewing the housing arrangements. In a small combined house and run system, houses with an upstairs sleeping and nest box area have the advantage of allowing birds to get out of each others' way. This also permits a better view of the birds' behaviour, and the high perch in the sleeping area makes inspecting them for damage easier. Canvas saddles can protect vulnerable breeding hens. However, feather pecking and cannibalism have differing causes, which can be difficult to comprehend.

▼ *Poultry will often be found enjoying a communal sunbathe or dust bath.*

▼ *Feather pecking may have complex causes and triggers.*

HANDLING POULTRY

While behaviour and the readiness of poultry to be handled may vary from breed to breed, it is the way that they are treated when chicks that most affects how poultry view humans, as well as how relaxed they are when they are handled.

Breeds vary significantly in temperament, from quiet and calm heavy breeds to flighty light breeds such as Hamburgs and Anconas. Hybrid strains that have been selected for their calmness in industrial situations may be the quietest and easiest to manage of all poultry. The nature of the breed determines how easy poultry are to handle, and so it follows that those new to poultry keeping would be best advised to choose a breed that doesn't mind, or even enjoys, being handled.

Nearly all the exhibition strains of large and bantam heavy, soft-feather breeds are quiet by nature and easy to handle. It is far easier to manage a flock of reasonably docile fowl than one of active fowl. It is also easier to treat an ailing individual if the bird is used to being handled.

▼ *Game fowl are among the most active, but remain calm when held at the "point of balance" by an experienced handler.*

Not all breeds or strains cope well with being handled. Fanciers keeping ancient populations of poultry breeds that are capable of living in naturally challenging situations should expect their flock to be extremely active and wary of being handled. These are birds that are best observed from a distance. Breeds such as Derbyshire Redcap, for instance, which are

▶ *A younger bird with its thighs gently but firmly held and supported on a forearm can be turned to be examined.*

PICKING UP FOWL

Handling poultry starts with knowing how to catch a hen, pick it up and hold it. Most egg-laying hybrids kept by professional producers are so quiet and tame that they will walk over your feet. Even with such calm creatures it is best to wait until a group of them are relaxing on an easily assessable perch before attempting to hold one, preferably at dusk or in subdued light.

Since airborne attack has always been a major threat to poultry, they are always far more at ease when approached at their level, rather than having a head and shoulders suddenly loom over them, and a pair of hands make a quick grab. With the growing confidence of the handler, the birds will accept being handled and it will soon be easy to examine the fowl in detail.

▲ *Approach the bird quietly when it is settled. Do not loom over it. Gently run your hands over the feathers towards the tail as if stroking a cat. Gently gather the wings around the bird's body.*

▲ *Gather the bird in your hands with one hand high on its thighs and the other thumb in the middle of its back. Gather the far wing around the bird, then gently lift her off the perch, possibly cradling her near side against your chest.*

▲ *Experienced breeders may appear to pick up day-old chicks by the handful, but the beginner would be best to gently gather them in both hands.*

evolved in a form that still enables them to look after themselves in challenging environments, will not welcome a high degree of human contact. Even exhibition breeds of

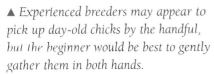

▼ *Older fowl used to being handled and loved by an owner will actually enjoy human contact.*

flighty birds that are used to expert handling will display a degree of activity in the show pen.

Handling chicks

All farmed birds are affected by the conditions in which they are kept and reared. Chicks, particularly bantams, brought up under a hen that has been handled from an early age will, when handled from one day old, soon learn to feed out of their owner's hand. Chicks hatched in an incubator will soon be accustomed to handling.

▲ *Very small children who love to handle chicks will need to be cautioned about grasping them too tightly, and are best encouraged to allow baby chicks to walk on to an open hand. In the same way, older children are best introduced to poultry by standing still and letting fowl walk around their feet, perhaps initially encouraged by a handful of grain.*

Hybrid fowl that have been selected for their calm behaviour may take months of careful handling before becoming even reasonably tame.

Removing and replacing a hen from a cage

1 A reasonably quiet hen can be gently turned to face an open door and then eased out of its pen.

2 The hen can be returned to the cage after being handled if allowed to walk in rather than being pushed.

FEEDING POULTRY

Like humans, poultry require a nutritionally balanced diet if they are to remain in good health and produce good quality eggs with golden yolks and strong shells. How and what to feed your poultry is a fundamental part of what every poultry keeper needs to know.

In the wild, poultry are sustained by a high-protein insect diet, as well as by the seeds, berries and foliage for which they forage. This comprises a rich source of nutrients that they peck at the ground to locate, or nip from the hedgerows. Wild birds will never consume to excess, and stop eating as soon as they have a full nutritional balance. Poultry only eat in daylight, so they need to satisfy all their food requirements in that time. In the summer, when there are long daylight hours, this is relatively easy since they have plenty of time to feed. In winter, however, some poultry can struggle to access all the nutrients they need to sustain them through short daytime hours and the night.

Vegetation

Grass contains essential nutrients that will vary according to the season as well as how often a run can be moved. Poultry that are free to range

▲ *The brassica family, which includes cabbage, cauliflower, broccoli and turnip, provides valuable nutrition.*

have the opportunity to find a proportion of their own food. Grass is essential to their diet and it is almost always possible to provide, especially where a movable run is employed.

For poultry confined to runs, hang a whole vegetable on a length of strong string and tie it at just above the birds' head height. The birds will

▲ *A balanced diet can be provided in pellet or mash form. Pellets are digested faster and are ideal for winter.*

jump to peck at the vegetable, providing them with a valuable distraction and thus preventing boredom. However, if you provide such a foodstuff and suddenly withdraw it, you will leave a gap in the birds' routine, and it may lead to problems of feather pecking.

Formulated poultry feed

The amount of feed consumed will vary enormously according to the size of bird and the amount of natural material it is able to find. A large fowl may consume 125g/4½oz of feed a day; some large exhibition fowl can eat far more. At the other end of the scale, the smallest Serama may eat less than 25g/1oz a day. While grass and vegetation add much to the diet and general wellbeing of poultry, bulkier supplementary food products

◄ *In the spring, grass can comprise more than 20 per cent protein for the poultry. Its high carotene content helps to create an egg yolk of good quality with a strong yellow colour.*

are necessary in all feeding regimes. Formulated poultry food needs to be kept in hoppers or troughs under cover so that the contents do not get wet. Unless consumed quickly, wet food will inevitably become smelly and unsavoury.

Complete layers' rations, as proprietary feeds are known, are cereal based and contain between 16 and 18 per cent protein, most of which is derived from soya bean products. These foods are appropriate to give to birds that are about to come into lay or are already laying eggs.

All the complete feeds intended for adult fowl are available either ground into meal or compressed into small pellets. The latter is almost dust free and is less wasteful than meal if spilt.

Mash is also available, a soft meal similar to ground-up pellets, but with a slightly lower nutritional value. This is because mash is usually fed to intensively housed birds and is thus available to eat from a hopper all day long. Since poultry only eat the amount of food that their bodies require, they need to keep eating mash in order to get their essential quota of nutrients. The theory is that if poultry spend more time eating they are less inclined to indulge in other, more antisocial, behaviour. Since intensively housed birds are likely to live in artificial light for 16 or 17 hours per day, they have sufficient time to eat their way through the day and get their full nutrient requirement. Mash is not suitable for birds that are free to roam and are able to forage for their own food, unless they are ex-battery hens used to a diet of mash. In this case, they should be fed higher-quality food mixed with the mash until such time as the mash can be removed.

Birds fed on pellets are able to consume their daily allowance in far less time than those fed on meal or dry mash. This is an advantage during the short winter days, when birds living according to natural daylight hours have less time to take in all the nutrients that they require to keep them going through the dark hours. During the longer, bright summer days, however, birds that eat quickly, particularly those that are semi-intensively housed birds, have time to become bored and start feather pecking and bullying.

POULTRY FEED
Nearly all of the available proprietary laying feeds, meaning feeds for egg-layers, that are available are based on years of research into the nutritional needs of the commercial laying hen, and have been designed to take care of their requirements without the addition of any other feed. All are cereal-based, with the addition of protein.

▲ *Pellets consist of compressed mash. This form is less likely to be blown away or lost when spilled.*

▲ *Mash or meal consists of ground cereals and other nutrients mixed together to provide a complete and balanced diet.*

▲ *Chick crumbs consist of specially formulated feed that has first been made into pellets and then ground to optimum size for chicks to eat.*

▲ *Mixed corn contains whole mixed wheat and broken maize grains. It is high in energy but lacks nutrients.*

▲ *Wheat may be purchased direct from a farm, but lacks the very high energy quotient given by added maize.*

▲ *Grit helps digestion and should be made available to fowl unlikely to access a natural source.*

▲ *Oyster shell assists digestion and can provide calcium, which is necessary to create healthy eggshells.*

▲ *Trough and drinking space will need to be increased as the flock grow.*

One solution could be to change to mash or a meal formula that takes longer to consume but will lack some nutrients. In the winter, poultry should get the food they need quickly, while in summer, providing feed of a lower nutritional value ensures that they eat for longer and have less opportunity to become bored.

Adding any extra grain to a formulated feed will alter the carefully calculated optimum nutritional value of it and change the ratio of protein to starch of the feed. Birds fed such a low-nutrition diet, may have their diet supplemented with a higher-energy feed containing wheat or maize. This is best fed in the late afternoon in small quantities, as a treat to help bring the birds back to the roost and also to provide them with something to sustain them during the night.

Mixed corn feed consists of whole wheat and a lower proportion of kibbled or very coarsely ground maize. Maize has higher energy levels than wheat and was traditionally used in very cold weather. When added to make up part of a laying hen's diet, it is found to enhance the colour of egg yolk. Some outlets may supply maize in an unmixed form, and farmers sometimes sell wheat at low prices. Feed merchants sell mixed corn feed and may offer maize to those who want to use it as a winter supplement.

Feeding the heavier traditional breeds extra wheat or mixed corn can help to satisfy their large appetites. Some strains of the old ultra-lightweight breeds that travel far in the search of protein-rich insects could also be fed extra grain to compensate for the energy they expend in their search.

Oats are no longer used in balanced feeds. These were once fed to the white-fleshed Sussex breed of table fowl that used to command a premium price at early London poultry markets. Oats were also used as a tonic in the feed of slower-developing exhibition breeds; the specific requirements of exhibition fowl may include diets rich in vitamins that enhance leg or feather colour, for example. These are available from specialist suppliers.

Grit and oyster shell

Hens that are free to range over a large area will seldom require extra grit to supplement their diet, since they can collect small stones from the garden environment. Coarse grit ensures a healthy digestive system, particularly when birds are free to consume long grass. It is accepted as good practice to provide grit in a small hopper, along with a separate hopper of oyster shell to maintain a regular supply. Oyster shell contains a source of calcium, which helps to promote strong eggshells. Most balanced mash and pellet feeds contain adequate amounts of this element, so calcium-rich oyster shell in ready supply will only be consumed as and when needed.

Water

Providing water is an essential aspect of poultry management. Fresh water needs to be available all day. It can be supplied through mains-fed systems or movable water founts.

▼ *Feed trough height is important and may need adjusting as birds grow.*

▼ *Fowl reared on pasture consume good quality grass, earthworms and insects.*

▲ *The bottom half of an egg box makes a useful and disposable feeder for day-old chicks.*

Chick feed

Poultry feed, particularly chick food, is expensive, although it costs no more to keep and rear well-bred and healthy stock than badly bred and ailing ones. It is always worth taking the time and spending a little extra to ensure a healthy foundation for the lifespan of poultry as well as for future generations of useful and productive fowl

Pullet feed

For traditional laying breeds that do not lay eggs much before 21 weeks, lower-protein "growers" feeds are specifically designed to delay laying until pullets are fully grown. Today's smaller hybrids are usually intensively reared in an environment where controlled lighting rather than food can be used to delay laying until the pullets reach 17 weeks. However, some of the more precocious strains will insist on laying before this age, and it may be necessary to feed them a specially formulated higher-protein "early-lay" formulation, available from feed merchants.

FOOD PROBLEMS

Boiling up household waste that may contain meat residues and giving it to poultry to eat is illegal. Providing poultry with boiled vegetarian kitchen scraps can also be harmful. Today's highly efficient hybrid probably lays 25 per cent more eggs, and weighs and has the capacity to eat 25 per cent less food than the pure-bred fowl of yesteryear. Free-range birds kept will come to little harm when consuming modest amounts of strange bits and pieces while roaming around the garden. However, the same fowl if kept in an enclosed run, will make themselves ill if they consume totally inappropriate food that they are unused to digesting. Feeding unsuitable bulky feed causes stomach upsets and diarrhoea.

Lawn trimmings from mowing and large quantities of earthworms and slugs mingled with the feed of hens that are unused to this type of food can also cause dietary issues. Grazed grass, however, is not a problem, since it is consumed in smaller quantities.

▲ *The divisions in this feeder stop birds scratching the food out.*

▲ *Water dispensers should be kept clean and filled regularly.*

▲ *An open trough has its uses, but allows smaller fowl to use feed as a dust bath.*

▼ *The divisions in this chick feeder stop food being scratched out on to the ground.*

INTRODUCING NEW BIRDS TO A FLOCK

Introducing new birds to an existing flock can often cause welfare problems, and so whenever possible, this is best avoided. The poultry pecking order that has been established will obviously be disturbed by a new arrival to a stable group.

Members of an existing flock are likely to bully a newcomer, so it is probably not worth trying to introduce a single bird to an existing flock. When introducing a small flock to an existing flock, the house to which they are introduced should be large enough to allow space for a netting partition between the groups until they have got used to each other. Even then, the two groups would probably mix better if both could be moved to a new house.

When moving birds into a flock that has an extensive run, a training run can be used for a few days until the newcomers get used to their surroundings, the long-term residents accept the new birds and the flocks get used to sharing their run.

A hen returning to the coop from brooding duties could spend a day in a temporary run before being placed on the perch in the main poultry house at night. The same technique is worth trying when adding a new bird

▲ *Hens that are new to an established poultry house can quickly become victims of bullying by senior hens.*

to an extensive run system. Such a bird may at first be unsure of the new surroundings and have to be encouraged to leave the security of the house. The bird in question would probably benefit from being

▼ *Poultry naturally choose to live in small flocks, and quickly establish a pecking order within that group.*

▲ *A recently mixed group of modern hybrid hens settling down together on a perch.*

removed from the run for a couple of mornings to spend another day or so in an acclimatization pen.

The optimum time to move birds to a new home is in the evening. This also allows you to see if the birds have become used to perching, or if they have been reared on a slatted floor system. Arriving at dusk will allow the birds to move on to a perch just as it is getting dark. At this point, birds will often be reassured by having their tails stroked. If the birds are going to use an outside run, keep them shut in for at least one day. When letting them out for the first time delay their exit until just before dark, so that they have enough time to familiarize themselves with the house entrance without straying too far. If the house has fitted nest boxes, deny the birds night access to these until they have learned to use perches, as once they start sleeping in these it is a difficult habit to break. In a perfect world, new birds would arrive before a weekend, providing you with time to observe their behaviour.

TRANSPORTING POULTRY

Restrictions imposed on the commercial transportation of livestock may have been, in part, responsible for the demise of some specialist pure-bred poultry flocks. More recently it may have led to the growth of poultry auctions.

Birds from specialist breeders were once efficiently transported by rail. Such poultry invariably arrived in good order because of the relatively airy carriages in use at that time. With the demise of this service, birds were sent by road instead, but rarely arrived in such good condition, and so this service was discontinued on welfare grounds. Today, other than the movement of small numbers of one's own birds, a licence and training are required before poultry can be transported more than 40 miles.

Birds that travel in cars, in properly ventilated boxes, in a temperature comfortable for the human occupants, are likely to be healthy on arrival. However, transporting birds in ventilated steel vans on modestly warm days can lead to heat distress. Traffic delays can lead to the death of such birds. In cold weather, poultry

▼ *Assess the weather before moving birds in an open-sided vehicle or trailer.*

need to be packed sufficiently closely together to keep warm, but still have adequate ventilation.

Carrying cases

Dealers may use plastic or wooden poultry crates; exhibitors may favour expensive wicker baskets. Plastic crates allow good air flow in an enclosed and ventilated vehicle, but are unsuitable for an open-sided trailer. A well-ventilated, strong cardboard box can be ideal when moving a few fowl, and is satisfactory in a closed van in cool conditions, but is not suitable on a hot day.

Three or four is the optimum number of birds likely to fit on the back seat of the average car in a carrying case. The box needs to be appropriate to the size of the bird – usually 45cm/15in square for bantams, and sufficiently tall for large fowl. Birds quickly settle in a darkened single or partitioned travelling box. As long as they

▲ *An exhibitor's bantam show hamper. Hessian lining and wicker construction allows natural ventilation.*

are not cramped and have room to stand, fowl that are used to spending up to ten hours on a perch do not need room to move about in transit. A 5 x 10cm/2 x 4in hole cut into two opposite sides of the box make perfect carrying handholds.

▼ *A show hamper suitably partitioned for two large fowl. Huge, fluffy exhibition fowl will require far larger hampers.*

PESTS AND DISEASES

Livestock health is a major concern for all those who keep poultry, and maintaining that health affects every aspect of their management and breeding. As poultry have the potential to suffer from as wide a range of health problems as any other livestock, these notes offer guidance.

Poultry occasionally suffer from a wider range of less common health problems than other domestic livestock, and this could put off potential keepers, but given the correct care, most fowl are as trouble-free as any other pet or free-range livestock. In many cases, careful observation of poultry can result in many potential problems being spotted at an early stage. This is easiest when birds are either in a large

▲ *A clean, healthy red face, with round, clear eye and nostrils free of clotted mucus, are indications of health and freedom from respiratory troubles.*

run or if the owner can walk into the poultry house. Stockmanship is about recognizing the difference between a seriously ill bird and one that is under the weather. Poultry keepers should soon recognize the difference between annual moult and feather pecking, as well as the difference between the usual degree of wear to the back of hens caused by an active male and the damage caused by the male's spurs or toenails, which may need trimming. Local poultry clubs may be able to demonstrate toenail, spur and beak trimming, and give

advice on what to do to save a visit to the local veterinary practice. Sometimes, fowl considered to be under the weather will perk up after a period under an infrared heat lamp.

To provide a yardstick for good health, it is easier if the first birds that a beginner keeps are as free as possible from inherited illness, as well as from health problems picked up from other fowl. Infections carried by stressed post-auction birds that have had minimal human contact may be severe. Birds used to being washed, handled and prepared for show, on the other hand, will return from show after show in full health.

Potential problems are likely to arise with the addition of new birds to a flock, when health problems that may be introduced by new birds are passed to the flock. Many poultry health problems faced by

▼ *A healthy, alert and well-feathered young fowl.*

non-commercial poultry keepers may be caused by mixing groups of young birds of differing health status. Others will be caused by an unwillingness on the part of the owner to dispose of ailing stock, or, in the case of ex-battery fowl, keeping prematurely aged fowl beyond what would be commercially viable. Observation and commonsense are paramount. Considering fowl as requiring a similar amount of care and attention as family members can also help.

Hybrid pullets born in commercial units will have been bred from disease-free parents. Reared in an enclosed bio-secure environment, they will have been subject to a vaccination program designed to give long-term protection against every disease known to attack commercial poultry. These include many respiratory diseases and salmonella that can make both the poultry, and occasionally those who consume their eggs, ill. Birds reared under less intense conditions rely on their acquired immunity to fight off infections.

▼ *A round, flesh-coloured vent through which the bird's faeces and eggs pass is one of the most important indicators of poultry health.*

In the case of ailments that require care and a degree of nurture, some means of isolating sick birds is essential. An isolation coop is a useful tool for dealing with the sort of problems that often see commercial poultry keepers deciding to dispose of ailing members of their flock.

The poultry tonics now approved for small-scale flock application may be kept on hand for use at times when poultry seem generally unwell.

Marek's disease

Poultry not routinely vaccinated against infection may be left susceptible to Marek's disease, a herpes-like infection that spreads through the respiratory system and can cause paralysis and ultimately the death of infected fowl. Chicks vaccinated against this virulent disease, however, may not pass their immunity on to the next generation. Some of the more attractive breeds

▼ *The high amounts of soya protein in modern poultry rations may produce faeces that are stickier than is ideal. A percentage of firm droppings with a white cap are a good indicator of health. Dietary upsets can cause diarrhoea, but a persistent runny discharge could indicate a worm infection.*

▲ *This soft-feather breed displays healthy feather regrowth following a period of moulting.*

remain susceptible to this painful and usually fatal illness, so owners may need to vaccinate each generation. It is always worth checking if a strain that you are purchasing is susceptible and has been vaccinated. While strains of some breeds are susceptible, others may be completely immune.

Mycoplasma

While many poultry health problems are interrelated, it is respiratory ailments that are likely to cause most concern. Poultry breathe through a series of air sacs and are consequently prone to respiratory trouble.

Chronic respiratory diseases (CRD) are often referred to as mycoplasma. Four main forms of mycoplasma are known, and they can be present singly in many birds without causing disease to exist. However, where combined with other viral and bacterial infections, they can result in outbreaks of sneezing, coughing and nasal discharge. Given poor ventilation in the poultry house or run, hygiene infections can spread rapidly, and while severity and

▲ *An out-of-condition hen that is going into a moult can be misdiagnosed as being sick. Poultry should be carefully monitored during any period of stress, and appropriate action taken.*

recovery times may vary, stress can contribute to high losses during some outbreaks of these.

Science has enabled the various strains of mycoplasma to be identified, but exact identification is difficult for the average poultry keeper. Tell-tale signs include runny nostrils and weepy eyes, combined in the early stages of infection with an uncomfortable flick of the head. If combined with infectious bronchitis (IB), which is impossible to identify without a laboratory, it is always likely to be labelled mycoplasma.

Mycoplasma are a genus of bacteria that do not have a cell wall. It is this genus that is also responsible for

causing pneumonia in humans. Prescribed treatments for other ailments break down illness by attacking the cell walls; however, since there are no cell walls,

▼ *Egg clusters or nits around the vent are often the first indication of this problem, which is easier to treat than an infestation of mites.*

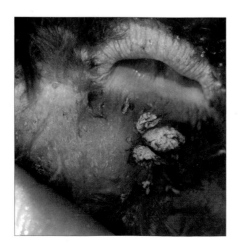

mycoplasma does not respond to traditional poultry tonics. In fact, there is no licensed vaccine for mycoplasma, although most of the larger hatcheries routinely vaccinate newly hatched chicks against IB.

Although viruses cannot be treated with antibiotics, in order to help birds cope with secondary infections, antibiotics may be routinely prescribed in mycoplasma outbreaks. The overuse of antibiotics is not good practice, however, and many veterinary practitioners may be reluctant to prescribe them. The best treatment is fresh air, clean dry litter and good management. A multi-vitamin preparation can make an effective treatment.

Bumble foot

This ailment is a swelling of the foot that can be caused by a thorn or sharp piece of grit becoming embedded in the soft underpad of the foot or toes. Attempts to lance the area of infection can make the problem worse. Less severe cases may be eased by keeping the bird on soft litter and soaking the foot in warm, salt water. Very heavy breeds may

▼ *Bumble foot may be caused by a wound becoming infected, or be the result of jumping off a high perch on to hard ground or stepping on thorns.*

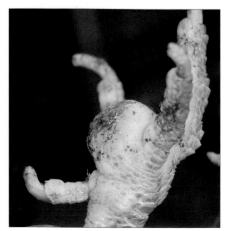

▲ *A prolapsed oviduct is most often a problem in older hens, but can be caused by strain during the laying process.*

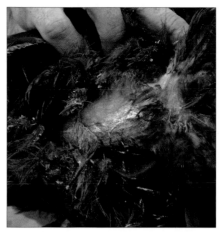

▲ *A foul odour and white discharge are the usual indication of vent gleet, which was once thought to be venereal disease, but is usually caused by ulceration of the vent and bowel.*

▲ *Diarrhoea is a condition that may arise from a variety of causes, including a sudden change of diet.*

suffer similar damage after jumping on to hard ground from perches that are too high. Check large, heavy breeds for bumblefoot as part of a routine health inspection.

Prolapsed oviduct

This condition may be prevalent in pullets coming into lay, or in old hens. The failure of the hen's muscles to support the oviduct under the strain of outsize eggs or heavy production, will result in reddish tissue protruding. Even mild cases of this condition can precipitate cannibalism, and so birds should be isolated as soon as possible. Treatment can be simple, but the problem can recur.

Dropped abdomen

A dropped abdomen is caused by muscle failure. Prolific layers and old, overweight birds are among those most often affected by this condition. While there is no treatment for this condition, birds may continue to lay but their eggs will rarely be fertile.

Vent gleet

Ulceration of the vent and lower bowel was at one time thought to be a venereal disease, but it is probably caused by bacteria entering the tissue. Vent gleet smells terrible. Many owners choose to dispose of affected

birds, but, at an early stage, isolation of the infected bird, followed by clipping the matted feathers from around the vent, then washing and treating the area with germicidal ointment that is suitable for humans can be effective if repeated over several days. Veterinary supervision will be required before antibiotics are prescribed in this situation.

Diarrhoea

From time to time, diarrhoea may be a problem for individual birds; however, temporary looseness of the bowels in individual birds should not necessarily be seen as evidence of contagious disease or infection. Diarrhoea may arise from a variety of causes, such as sudden changes in the feed regime or simply the wrong sort of food being provided. Too much airflow in slatted-floor houses can cause abdominal chills, which may trigger diarrhoea. Confined, stuffy houses have the same effect. Infected flocks are treated by a dose of Epsom salts (55g/2oz per 4.5 litres/1 gallon of drinking water). The inclusion of 2 per cent vegetable charcoal in the diet is a reliable treatment in mild cases of dietary upset.

TREATING A PROLAPSED OVIDUCT

Wash the vent with warm disinfectant solution and when clean, gently smear with a proprietary ointment. Gently replace the parts and syringe the vent with cold water to stimulate muscle reaction. A teaspoon of olive oil taken orally for a few days will help to discourage egg production during the recovery period.

▲ *A beak that is too large for an underdeveloped head can indicate an unsatisfactory start in life.*

Parasites

Internal and external parasites are a recurring occasional irritant to both poultry and those who care for them. At times they can escalate to become a serious problem. Poultry have always suffered from parasites, ranging from visible external lice and mites, to the less obvious internal worms, and outbreaks of protozoa, which cause the parasitic diseases coccidiosis and blackhead.

Worms

Domestic fowl are susceptible to infestations of several different types of intestinal worms, with round worms being the most common. While many fowl, particularly free-range birds, will shrug off minor infestations, and routine hygiene will limit outbreaks in intensively housed birds, modern poultry wormers are so safe and effective that most keepers will still opt for the security and peace of mind of an annual or biannual treatment program. Symptoms of worms include ruffled feathers, generally malaise, a bird

▲ *Flowers of sulphur was once used to treat lice and fleas. It can be an effective treatment in the early stages of leg mite. However, a liberal application of petroleum jelly may be just as effective.*

with a messy vent often accompanied by congealed faeces, and clumps of lice eggs. Stunted and uneven growth among young stock can suggest an ongoing problem. Appropriate modern worming treatments are available to eradicate the problem. The gapeworm, a nematode that infests the windpipes and once caused heavy losses in young poultry, can be treated in the same way.

Lice

Lice infestations, like intestinal worms, are often found in or on generally unhealthy fowl. Given access to a large run and a dry dust-bathing area, healthy fowl will rarely suffer from fleas or lice. Lice chew the flesh and feathers and cause irritation that affects the overall health of the fowl, so regular inspections should be made. Lice are light brown, flattish insects, and their tell-tale egg secretions will be found cemented to feather shafts. Lice are easily killed, usually by spraying or dusting with a

readily available proprietary product. Should re-infestations occur, treatment or at least a weekly inspection could become part of the hygiene routine.

Mites

Dealing with the various mites that plague poultry and their housing can be difficult. These are parasites that lives in the hen house rather than on the birds, they needs to be checked for regularly and treated as part of house cleaning and hygiene. The blood-sucking red mite lives in crevices in the house and feeds when the birds are on the perch. Infestations can be confined to perch ends for weeks, only to multiply in days to a point where they will make life unbearable for both occupants and people who have to go into the house. The problem is exacerbated in the summer when the tell-tale piles of dust at perch joints will be visible. These may develop almost overnight into a seething mass of blood-filled

▼ *The effect of scaly leg mite can be long-term disruption of the scales that make up the outer covering of birds' legs and feet. An application of petroleum jelly will help soften the scales, but it may take up to two years for the scales to be replaced with new ones.*

mites. Treatments kill the insects by rapidly lowering the moisture levels of their bodies.

The related northern or plumage mite is spread by direct contact. It can burrow deep into feather follicles, leaving characteristic curled or damaged feathers. Vulnerability seems to vary from breed and strain, but many exhibitors will isolate affected birds and in some cases dust or spray them with proprietary products. The effectiveness of these products varies.

The scaly leg mite can burrow deep into feet and leg scales, and can cause painful and unsightly disruption of the scales. This mite can be eradicated by insect sprays, provided that they can penetrate the area where the mite is, but further treatment will be needed to repair the scale damage. Repeated applications of petroleum jelly are one of the best treatments for the injured area. The jelly works by suffocating the mite and may also encourage re-growth of healthy scales.

Coccidiosis

A protozoan disease, coccidiosis can be a particular problem in young chicks. Its prevention should be considered as part of the rearing regime. It is caused by living micro scopic organisms (protozoa) called coccidea, different types of which are found in chickens and turkeys. The protozoa undergo a very complex cycle of development inside and outside the birds' bodies. It appears that birds reared outdoors may be more at risk of coccidiosis than those reared in more controlled situations.

Outbreaks will also often occur in groups of chicks being reared in isolation, either under lamps or in brooders. Here, any build-up of damp or dirty litter may, particularly in very hot weather, result in whole batches of young birds becoming ill, and may result in the death of untreated birds.

▲ *Knocks and blisters on the comb resulting from injuries or dried on dirt can usually be treated with petroleum jelly or household germicidal cream.*

Infected chicks present a dejected and ruffled appearance, often accompanied by pathetic cheeping. Bloody droppings are sufficient evidence of the presence of coccidiosis, and treatment is prescribed by a veterinary surgeon. Most treated birds are likely to recover and may acquire resulting immunity. Suitably medicated chick food will give a wide measure of protection until birds gain their own immunity. Some breeders have reported gaining some measure of control over outbreaks by the inclusion of cider vinegar in the drinking water.

Blackhead

Blackhead is a protozoan disease that infects turkeys and can have devastating consequences. It derives its name from the darkening of the head in some, but not all, affected birds. It is an acutely infectious disease common among turkeys, pea-fowl and related species. It has always been a problem in houses that are shared with other classes of poultry,

▲ *An inflamed crop can be an early indication of a crop binding, or sour crop which is frequently caused by birds consuming long grass. Relief is given by pouring a teaspoon of olive oil down the oesophagus. Removal of the obstruction may require the crop to be operated upon.*

since chickens carry, but are not necessarily affected by it. Blackhead is a problem that has increased as commercial production has become more intensive. The industry once relied on a group of powerful drugs for protection to combat the disease, incorporating them into the feed, or keeping them on hand in case of sudden outbreaks. Once it was discovered that residues of these drugs could pose a risk to humans, their use was banned in many countries, and turkey producers were concerned for the future of the industry. However, smaller and organic producers relied on a herbal preventive treatment based on oregano. This product has been so successful, involving only minor adjustments to the way in which turkeys are kept, that many people now question whether the industry has become too reliant on potentially dangerous medications. Turkey owners should be continually vigilant.

EGG PROBLEMS

A regular supply of fresh eggs is one of the main reasons for keeping poultry. Problems can occur with both the eggshells and their contents, however. This section describes some of the most common problems and offers advice on what can be done.

Occasionally, grotesquely misshapen or oversized eggs are laid that are best dismissed as a freak of nature. Sometimes the shell may be covered in blood. This is because the hen will have strained to lay the egg. It does not affect the quality of the content. Double-yoked eggs are comparatively common, the occasional triple-yolk ones less so. They may arise because two or three yolks leave the ovary and pass down the oviduct together or in rapid succession, ending up within the same shell. It is rare that a bird produces one egg inside another. If it does happen, then it is caused by a shock to the oviduct, which has caused one egg to move back up the oviduct until it meets another one coming down.

Eggshells

Shell quality and defects are immediately obvious. Soft-shelled, very thin-shelled eggs, or even shell-less eggs are often wrongly attributed to calcium deficiency. However, all

▼ *Crinkle and ripple-shelled eggs are caused by worn and stretched oviducts. Older high-yield hybrids are prone to this.*

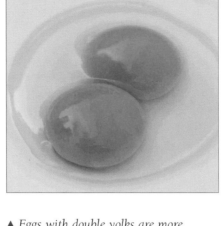

▲ *Eggs with double yolks are more likely to occur either when pullets are beginning to lay for the first time, or more rarely, when older hens are coming back into lay after a longish rest.*

proprietary poultry feed contains adequate calcium combined with the correct phosphorus ratio and optimum levels of vitamin D to produce eggs with hard shells. The problem is most likely to occur during rapid increases in egg production, when very young, high-

▼ *Eggs with rough-textured or "sandpaper" ends can be an indication of a mycoplasma infection.*

production hybrids, and occasionally older birds that haven't laid for a while, start coming into lay. Because the ovary is in an extremely active condition, yolks leave it in quick succession and the hens are unable to assimilate enough of the available calcium to provide shells for all of the eggs being produced. The problem usually disappears as the ovary becomes less active.

The reason why some strains of lighter coloured hybrids start laying nearly white rather than brown eggs

▼ *A "faded" eggshell compared with a standard egg, both of which are produced by the same breed.*

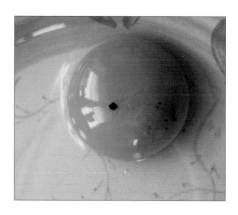

▲ *Soft-shelled eggs are generally laid by young hens coming into lay for the first time, or by older birds.*

▲ *An egg yolk that fails to form may be produced by a young hen coming into lay.*

▲ *A blood spot detracts from the appeal of an egg but is harmless and can be removed easily.*

during periods of intense sunlight is barely understood. The problem seems to be caused by ultraviolet light, haemoglobin and a little understood genetic defect that affects some strains with very pale brown feathers that lay brown eggs. Some poultry keepers observe that the problem seems worse during periods of normal summer feather loss. As this is genetic, there is no cure other than restricting access to full sun.

Inside the shell
Poor eggshells detract from both show quality and saleability, but internal defects are likely to concern

the consumer far more. Blood spots in both the yolk and the white are relatively common. The former is caused by an intra follicular haemorrhage while the ovum is still in the follicle. The latter is caused by the rupture of a blood vessel in the oviduct at the time that the yolk is passing through it. Blood spots vary in size. The problem is at its worst when hens are coming into lay. This is the time when egg classes at exhibitions are at their largest too, making the problem appear more prevalent. Eggs with small blood spots may be perfectly edible. The problem usually disappears when the

ovary becomes less active, and there is no cure. Meat spots are formed from blood spots that have degenerated after inclusion in the egg, and not, as was at one time thought, due to the inclusion of meat meal in the feed.

The cause of green and other miscoloured yolks is dietary. Hens produce wonderful, yellow yolks when they have access to grass. A slight green tinge may be associated with eating clover and other legumes. Olive and even black-yoked duck eggs are caused by the consumption of ripe acorns or using oak and some other hardwood shaving as litter in the hen house.

▼ *Sometimes an indication of mycoplasma, rough-shelled eggs are attributed to some shock to the hen.*

▼ *Dirty eggs need to be cleaned quickly since the shells are porous and the insides can become contaminated.*

▼ *Collapsed yolks can be the result of a dietary disorder, but can also be caused by rough handling after laying.*

BREEDING POULTRY

Once you've had success keeping poultry, the next stage for many owners may be to set up a breeding program to ensure a continuous supply of future generations of fowl. This is particularly the case if you have good stock and wish to have more of it, or if you think you can improve the utility or exhibition value of your chosen breed. For those interested in genetic conservation, the wish to add to the distinguished gene pool of an old or rare breed may be sufficient motivation to begin breeding poultry. In the latter cases, contact with the relevant breed club is essential. For many other poultry owners, including those who have children, the desire to breed poultry is sparked by the simple joy and satisfaction of watching a mother hen fussing over a new brood, or capturing the moment when new life emerges from an incubated egg. While putting a few eggs under a broody hen and hoping for the best can be fun, things do not always go to plan, and plenty of information is included in this chapter on the breeding process and on how to deal with the unexpected.

▲ *The sight of a mother hen and newly hatched chicks will enchant both adults and children.*

◄ *A mother hen nurtures her brood of growing chicks, just as nature intended.*

SETTING UP A BREEDING PROGRAM

To start a breeding program a healthy male and at least one healthy female is required. Most programs involve a male and two or more females; with some light-breed males managing to cope with eight or more females.

The less closely the male is related to the female the more vigour any offspring are likely to have, meaning that the resulting generation should be as free as possible of genetic faults. Provided the male is fertile, male and female birds will mate and a clutch of fertile eggs will result. Breeding poultry produce 50 per cent male and 50 per cent female offspring.

If you are considering breeding poultry, it is quite reasonable to expect that like will beget like. As part of the ethos behind the standardization of a pure breed, the Poultry Club of any country asks for evidence that subsequent generations will "breed true". This means that the offspring will be identical to the parents in terms of visual appearance,

▼ Poultry selected for breeding will be the birds with the best characteristics in the owner's stock.

character and temperament. While the emphasis on visual appearance remains the basis of exhibition breeding and selection, most utility selection, for egg and meat production, has always been based on specific genetic properties. Usual commercial practice involves crossing different breeds and strains to produce a generation that not only has hybrid vigour but also encapsulates the desirable qualities of both parent lines. This keeps the breeding stock healthy.

Pure breeds and hybrids

For birds that are not being bred specifically for exhibition, a male bird is usually allowed to run in a breeding pen with a handful of females. Both male and female birds in the pen will exhibit all the desirable traits for the breed. Several male birds may be required for a

▲ A young red or genetically gold male saved for use in a quality table fowl breeding program.

largish flock of females, with the result that it will not be possible to determine the exact male parent of the offspring.

Exhibition breeds

If you wish to breed birds for exhibition, always begin with the best stock that you have or can buy.

> **BIRDS THAT DON'T MAKE THE GRADE**
> Other than keeping males for breeding or for fattening for the table, most males will need to be disposed of as young as possible. Similarly, birds that do not exhibit sufficient characteristics to match the breed standard should be discarded.

It takes longer to build good stock from average birds than to breed a few undesirable traits out of quality stock. As most pure breeds descend from relatively few ancestors, breeding to type (creating offspring that match the parents) may result in having to breed poultry with parents that are distantly related. In fact, inbreeding is essential for pure breed lines, and it is common to cross offspring back to their parents or with their siblings. The result is that hybrid vigour is lost, but desirable traits are retained. Hybrid vigour can be reinstated by breeding unrelated birds from the same breed, but there is no guarantee that undesirable traits from the new birds' heritage will not resurface in the next generation. For exhibition, pure breeds are selectively bred, meaning that individual birds are chosen to breed for specific traits. The lineage is clearly traceable with exhibition birds, and each is tagged to aid identification.

Double mating

Generally applied to exhibition birds, the term double mating is used to describe the practice of selecting one breeding group to breed exhibition birds of one sex, and a second group

▲ A Blue Cochin mother and father produce Black, Blue and dilute blue, or Splashed blue chicks.

to breed the other sex. For example, many of the Mediterranean exhibition-class breeds have large single combs. In the exhibition male the comb is expected to stand boldly upright, while the female comb is expected to fold gracefully over to one side. In order to stand upright the tall male comb needs a wide strong base. In order to flop gracefully to one side the female comb needs to be relatively slim. Some breeders will make up a "cock breeder" group headed by a male with the required tall upright comb, along with hens that have a shorter, stronger comb than would be the exhibition ideal. The intention is to produce offspring with the desirable tall, upright combs that have a strong base. Similarly, "pullet breeder" groups consist of hens with fine, floppy exhibition combs, headed by a male with a large but less substantial comb than would be the exhibition ideal, in order to produce a large floppy comb.

GREGOR MENDEL

The 19th-century scientist, Gregor Mendel, found that by crossing two tall pea plants, the offspring (the F1 generation) were equally tall. When the F1 generation were crossed, however, only 50 per cent of the offspring (the F2 generation) had the combined characteristics of both parents, while 25 per cent had the characteristics of one parent, and 25 per cent had characteristics identical to the other.

The principles of this theory are used, for example, by breeders of the Blue Andalusian breed. In order to produce a Blue bird, a White male is crossed with a Black female. However, if the Blue offspring are mated, 50 per cent of the next generation will be Blue, 25 per cent per cent Black and 25 per cent pale

diluted blues or "Splashed". When Black and Splashed birds are bred together, 100 per cent of the resulting chicks feather with a blue-grey hue.

The Frizzle breed has broad backward-curving feathers. Pairing two Frizzles will produce offspring in which 50 per cent are identical to the parents, 25 per cent will have narrow, over-frizzled feathers, and 25 per cent regular, straight feathers. Although 100 per cent of the chicks are Frizzles, at least superficially, only half of the offspring breed true. Birds that do not breed true are discarded from any future breeding pool.

Many desirable exhibition-standard birds can only be achieved by pairings that will produce a given percentage of offspring that are different to the desired breed standard.

AUTO-SEXED AND SEX-LINKED BREEDS

The offspring of auto-sexed and sex-linked breeds can be identified as male and female by the colour of their down when they are one day old. The advantage for the egg-producing industry is that unproductive male birds can be discarded, keeping resources focused on the valuable hens.

In order to produce auto-sexed and sex-linked offspring, specific pure breeds must be used in the breeding programs.

Sex-linked breeding

Birds bred to produce chicks in which the offspring are easily distinguished by the colour of their down are planned crosses. Only certain pure breeds, when crossed with other specific pure breeds, produce offspring that can be sexed on hatching. As well as being able to identify male and female offspring, which is economically advantageous for the poultry industry, the offspring will be hardy and have a greater utility value than the breeds' parents. The crossing of two specific breeds will always result in crossbred offspring. The next generation

▼ *A typical sex-linked pairing is a genetically gold male such as a Rhode Island Red, crossed with a genetically silver female such as a Light Sussex.*

produced by these first crosses, like hybrids, can be sexed at one day old. The first generation of crosses can be used for breeding, but the colour, and to some extent the type (shape) of resulting generations will not be predictable. It is also impossible to distinguish whether the offspring is male or female at a day old, unlike in auto-sexing breeds.

▲ *The offspring of the Rhode Island Red and Light Sussex crossing produces male offspring with light yellow down and female offspring with brown down.*

Auto-sexing breeds

In 1920, on the basis of research by renowned geneticists Professor Punnett and his associate Michael Pease, a Barred Plymouth Rock was crossed with a Gold Campine to create offspring with a completely new type of plumage pattern known as Barring. The result of the cross was the Cambar breed, now known as an auto-sexing breed.

Auto-sexing is a method of identifying the sex of young birds at one day old so that unwanted male birds may be culled, thus reducing food and housing costs for those birds. Of those poultry breeds that are auto-sexing, it is the colour of the chicks' down that indicates whether the chick is male or female. The down may be light for one sex and dark for the other, or one sex may be identified by clear spots or barring.

▶ *These chicks are one day old and have been bred so that males and females can be identified easily by the significant difference in down colour.*

Additionally, with an auto-sexed pairing, the resultant offspring is always a pure breed, meaning that if the new Cambar offspring are mated, the third and subsequent generations will always breed true to the Cambar breed.

An understanding of auto-sexing breeds was a significant commercial development prior to the arrival of the hybrid lines. This research showed that science could be used to modify poultry breeds and for the first time, birds could be bred to produce chicks that were effectively colour-coded on hatching, with the colour of the down indicating which were male and which were female. Breeders soon found that even when Barred males were mated with females of other breeds, it was still possible to tell the sex of the offspring upon hatching. Later it was found that other barred breeds could pass that characteristic on to other auto-sexing breeds.

The group of breeds whose names

end in "-bar" are auto-sexing varieties of a parent breed that has had barring added to its genes in such a way that generation after generation of their offspring can be feather-sexed by their colour at one day old.

The other half of the breed's descriptive name, such as "Wel" in Welbar, for example, is from one parent of the new breed, the pure breed Welsummer, in this case. Other auto-sexing breeds or varieties follow the same formula, the first part of the name denoting the parent breed, with "bar" added to show this modification. However, it should be pointed out that in spite of their

▼ *The Cream, or Crested, Legbar was created using Araucana rather than Leghorn. Its name is derived from the down colour of the chicks, but breaks the rules governing the naming of the other auto-sexing breeds or variants.*

adopted names, many hybrid fowl that have the suffix "bar" do not auto-sex.

Another development is that bantam versions of many auto-sexing breeds have been created using the same formula, and, like other auto-sexing breeds, other than the barring factor they are expected to be otherwise the same as their original unbarred parent breed. Their essential characteristics will be expected to exactly replicate those of that parent.

▼ *The natural barring found in the newly created Suffolk Chequers would provide a basic feather pattern for any auto-sexing project.*

Crossing two barred breeds

The barring of the plumage in auto-sexing breeds is finer and more sharply defined than that of the Pencilled pure breeds. The auto-sexed Barred Rock, which has black down with a white head patch, was crossed with a pure breed (Barred) Gold Campine, which has mottled brown down with a light head patch, to determine the offspring's plumage.

As the chicks hatched, only those that showed the light head patch and brown down (of the Gold Campine), which combined the dominant sex-linked barring with the recessive pure breed Campine barring did not auto-sex. These birds were heterozygous (impure) for sex-linked barring. The second generation, however, when mated, gave a combination with homozygous (pure) sex-linked barring, and an entirely new down type appeared. This new down was strikingly paler. The light head patch had spread over the neck and back, blurring the sharp, spotty pattern of the Campine down. These chicks were males. This new breed was named Cambar to reflect its parentage.

The barring factor was so complete that given a proper understanding of the genetic formula, a breeder could increase the content of the original pure breed to the point that it was at least 90 per cent pure for that breed.

This finding opened the way to create a range of auto-sexing forms of both well-known commercial breeds.

As light breeds, Leghorns are among the best layers. A new-found ability to eliminate the males, which have no table value, at hatching meant that the Gold Legbar variety had for a while considerable commercial value. The obscure Araucana was used to create the Crested or Cream Legbar for what was potentially the best producer of blue eggs. Its descendants still have a commercial role to this day.

Such pioneering work was only possible because small-scale hobby breeders refined their exhibition breeds to have the purity of feather pattern which was expected in a standardized breed.

▼ *The Wybar bantam was created from a Barred and Silver-laced Wyandotte.*

Welbar

The Welbar is a British bird created in the 1940s from Gold Barred Plymouth Rock hens crossed with Silver Welsummer males. The "Wel" in the name refers to the Welsummer male's genetic input.

This utility breed was primarily bred to lay plenty of dark brown eggs, and the feather colour of the breed remains of secondary value. The male is gold with black barring. Hens are paler versions of the same colour. Just one variety exists now, although two were developed. The silver variety would be easy to recreate, however.

The Welbar has the broad carcass of all egg-laying breeds, with a long and deep breast. This is an attractive and upright bird that holds its tail

▲ *The Welbar bantam is a Welsummer bantam with an added barring factor that enables feather sexing of its day-old chicks.*

▼ *A large version of the German auto-sexing Bielefelder breed.*

aloft. It has a neat habit with soft, silky plumage, and the yellow legs should be clear of feathering.

Wybar

Like the Welbar, the Wybar uses Barred Plymouth Rock in its composition, this time crossed with Silver-laced Wyandotte. It is a British fowl developed in 1950 to be a dual-purpose bird with exhibition qualities. It is likely that it also has Light Sussex added to its genetic pool, though the bird predominantly takes its mother's characteristics. It has a pea comb, silver-laced coloration, yellow legs and beak. The basic formula can still be used to create new varieties, including bantams. Wybar fanciers have created a small fowl that enables breeders to eliminate unwanted male chicks at hatching.

Bielefelder

A German breed created out of a broadly barred form of Plymouth Rock and Rhode Island and New Hampshire Reds, the Bielefelder is an attractive heavy breed with chicks that can be sexed at one day old by down colour. This is a relatively new breed developed in the 1970s. Bielefelder are available in just one colourway, with males being paler versions of the hens. It was bred as an egg-laying breed that was tolerant of cold.

Suffolk Chequers

Not yet standardized, the Suffolk Chequers has been created very much in the form of earlier Plymouth Rocks in which broader feathers enabled their pure-bred offspring to be feather-sexed at one day old.

Rhodebar

The Rhodebar was created in Britain in the 1900s. It was produced by crossing Barred Plymouth Rock males, an auto-sexing breed, with Rhode Island Red hens. The result was originally called a Redbar but later renamed Rhodebar. The breed was created as an egg producer, with hens laying up to 200 brown-tinted eggs in their first season. Adult Rhodebars look similar to Rhode Island Reds, but have a black tip to the tail. Day-old male chicks are yellow, while females have dark stripes or barring down their backs.

◄ *Suffolk Chequers is a new breed in the UK and is yet to be approved by the Poultry Club of Great Britain.*

▶ *The Rhodebar has a large, deep, broad body and a long back. The single comb, earlobes and wattles are red, and the tail has a black tip.*

REARING CHICKS UNDER A BROODY HEN

Many hen breeds have a tendency to broodiness. This is an inclination to sit on the eggs, to allow them to hatch, and then to nurture a young brood of chicks. Hens will do this regardless of whether they have mated with a rooster.

Some breeds are more prone to broodiness than others, and may become broody at any time of year. Other fowl are most likely to be broody and reach the peak laying period when the daylight hours lengthen in spring. Broodiness is not always a welcome characteristic and retrieving eggs, particularly unfertilized eggs, from under a broody hen can be difficult.

The broody hen

A hen that is broody during the autumn and winter, when daylight hours are shortest, produces chicks that inevitably get off to a slower start than their artificially incubated counterparts. Some breeds have had the brooding instinct bred out of them, particularly breeds that are

▼ The enclosed apex ark makes a cosy place for a hen to incubate a clutch of eggs, but a rapidly growing brood can quickly outgrow the space.

▲ A broody hen has a raised body temperature. She will stay on the nest at night and tuck every available egg under her protective skirt of feathers.

egg-layers. Broody hens can alter the balance of the pecking order, and other birds in the flock may become aggressive towards a broody hen.

If you have a cockerel running with the hens, unless you want chicks, collect the eggs every day. A hen will lay roughly one egg per day and will sit on them only when she has her full clutch. A clutch could be up to 12 eggs. The hen waits for a full clutch before she sits on them so that the chicks hatch at the same time. A hen that sits on her eggs and does not move for several hours may be broody. She may also be disinclined to leave the nesting box, and may start nesting. A broody hen's body becomes hotter than usual, and she may remove feathers from her chest area so that the eggs benefit from the increased body heat. The hen will sit on her eggs, spreading her wings over

▲ Large fowl chicks may quickly outgrow a surrogate bantam's capacity to nurture in cold weather, but these chicks will soon be sufficiently feathered to keep warm.

them for 21 days, only leaving them once a day to tend to her own needs. Each day she will turn the eggs to ensure the embryos develop evenly.

Broody hens need continual assessment to ensure that they are not neglecting their own health. Not all hens that have sat tightly on their eggs can be relied upon to be trouble-free mothers.

Some breeders think that chicks hatched under a hen in their natural environment have a healthier start in life than those reared in arguably more sterile conditions under the continual warmth of an infrared heat lamp. Hens that hatch and rear their own chicks may be naturally stronger. Without human intervention only the fittest survive; this hardiness is often a desirable trait in specific breeds. Chicks brought up by hens often display very different characteristics

▲ *Big breeds like Cochins and Brahmas are natural broodies that can cover a lot of eggs, but their large size and feathered feet make some hens clumsy mothers.*

from those hatched in an incubator, where the first glimpse of another living creature may be the person that handles it and on whom it is dependent for food and water. Chicks hatched under hens rely on the hen to call them back to the coop. They have the additional security of being able to hide under their mother's feathers. If they are not regularly handled during this period, they may grow up with a natural suspicion of humans. Yet chicks brooded under the tamest bantam hens, which have also been bred and reared in tiny pens, often at head and shoulder level with the sort of breeder who constantly checks on their progress, can soon be found walking up an outstretched forearm in order to get a better look at the benefactor who comes several times a day to tend to their every need.

Discouraging broodiness

Stopping a hen from broody behaviour is possible and is most easily done as soon as the behaviour commences. Hens allowed to be persistently broody can, after some weeks, become debilitated. Putting the hen in a coop that has a wire mesh base, which is set upon bricks at its edges, makes an uncomfortable surface for the hen, so that she will stand rather than attempt to nest. The cold air circulating from below also helps to cool her body temperature. A few days spent on draughty slats should cure a broody hen and is a short-term solution that a battery hen endures all its life.

Another method is to fill the nest box in the coop with an obstacle that the hen is unable to displace. This will encourage the hen to move on.

▶ *A mother hen will call her chicks to bring them home. She will scold and issue alarm calls on the approach of unfamiliar or potentially threatening humans or animals.*

Housing for mother and chicks

A mother hen carefully brooding her family of chicks is an appealing sight. Her natural maternal instincts are best served by simple but specialized housing. Left to her own devices, a mother hen may drag her chicks too far in the search for food. The broody coop, which was used when much of the industry still relied on natural methods, provides a safe, dry home for the mother hen, while allowing chicks access to fresh air and their own chick feeders. While they are suitably confined with their mother, chicks may choose not to venture too far, at least for a few days. Cats and other threats will not be able to get too close while the chicks are kept within a small, safe enclosure.

A well-designed coop allows easy access to the hen through a sliding roof panel or removable front rails. These rails confine the hen and chicks to the coop, but allow enough space for them to live comfortably. Confining the hen after hatching, while allowing the chicks plenty of room to exercise in the protected run, ensures the hen does not take the chicks too far. If day-old chicks are free to roam it can result in loss.

HATCHING USING AN INCUBATOR

Artificial incubation is a method to bring fertile eggs safely from point of lay to hatching without including the hen in the process. This procedure was initially developed to increase the volume of eggs hatched by breeds that were less likely to go broody.

The process of artificial incubation mimics the natural function of the broody hen sitting on her eggs, keeping them warm and turning them regularly until the point of hatching.

Artificial brooding methods

The industry has relied increasingly on artificial incubation since the 1930s. The early "brooders", or incubators, sought to mimic the dark and cosy warmth provided by the hen. Early "foster mothers", as they were known, relied on paraffin to generate the heat and consisted of an enclosed oil lamp under an insulated canopy, surrounded by a felt skirt. It provided similar insulation to the sitting hen's feathers. The eggs were turned by hand.

Fertile eggs could be hatched earlier in the year than had previously been possible. A lack of daylight, however, meant the newly hatched chicks spent longer staying close to

◄ *The tiny low-voltage incubator enables schools and other organizations to safely demonstrate the miracle of hatching to groups of young children.*

▲ *The incubation process extends the laying season beyond that in which hens naturally hatch and brood chicks.*

the warm glow of their brooder lamp than they did eating and drinking, and so some early batches failed to benefit from the head start that their early hatching should have offered. Apart from being a bit smaller than those chicks bred naturally at a more usual time of year few of those chicks in fact came to harm, provided a reasonable room temperature could be maintained. Sunlight was thought important to the chicks' development and as this was readily available in the form of vitamin D it was included with cod liver oil in the chicks' feed.

Contemporary methods

The advent of reliable miniature appliances, like the tiny seven-egg incubator, has enabled eggs to be

▲ *Incubators are available in a range of sizes, from mini seven-egg models to designs housing thousands of eggs.*

hatched at home. Ideally, a small group of eggs laid within a day or so of each other will be put in the incubator at the same time. Being genetically similar, they should hatch within an hour or so of each other. Incubators are available in different sizes.

The incubation process from point of lay to hatching can take up to 21 days, with some chicks hatching at 18 days and others taking a few days longer. Any eggs that have not hatched after this time should be discarded, since retaining them can lead to disease. Modern incubators are run digitally, and enable egg temperature to be controlled to within a fraction of the optimum temperature of 37.5°C. Even the smallest machines now available for home use have efficient self-turning systems, which means that poor hatches, particularly those resulting in excessive numbers of chicks dead in the shell, should practically be

Hatching

1 Hatching starts with the chick pecking, or "pipping", a tiny hole in the eggshell around the air sac.

2 The first tiny holes appear at the point of the enlarged air space in the blunt end of the egg.

3 The chick may not do much more for an hour, or it may quickly peck its way around the edge of the air cavity.

4 The chick pauses to take its first breath of open air. Then with one heave it can be almost free of the shell.

5 Some chicks take a little longer to break free of the egg shell, but one should never be tempted to help speed up the process.

6 As the chick emerges it is possible to observe the "egg tooth" on top of the upper mandible that the chick has used to hammer through the shell.

eliminated. Breeders planning larger hatching operations could look at a bigger incubator system, one in which the eggs are removed for the last two or three days of the incubation period to a separate hatching compartment, or better still, to a completely different device known as a "hatcher".

If the embryo is not to drown during the latter stages of development, the egg will need to lose moisture during incubation. If turned properly, a healthy developing embryo should hatch into a healthy, fluffy chick even without perfect conditions. After all, hens that sit on eggs are not always able to provide ideal conditions themselves.

Even the smallest incubators are fitted with some method of controlling moisture and humidity, with 60% being the optimum temperature. The incubator should remain closed during hatching in order to retain moisture levels. Experienced breeders may have

▶ *Empty egg shells with a few unhatched eggs are indications of a reasonable hatch in a conventional incubator.*

▲ *A few hours old, and this healthy batch of Legbar chicks are nicely dried and ready for transfer to a brooder unit.*

differing opinions about the moisture levels. Some of the more successful breeders, when using modern fan-assisted incubators, add little extra water during the early stages of incubation, only increasing moisture during the hatching period. Slow, late or difficult hatches, particularly those where nearly all of the emerging chicks have difficulty getting out of the shell and appear to be dry and sticky, are often wrongly blamed on lack of moisture. The viscous excess fluid on the chicks dries like glue

▼ *By comparing the internal view of the egg to a candling chart, it is possible to see if the egg is fertile or not.*

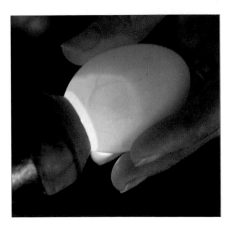

when in contact with air that enters the shell as the chick "pips" or begins to break out.

The incubation period will vary between breeds and last longer when eggs have been stored before being placed in the incubator. Hatching may start as early as the 18th day when a chip appears along the line of the air space. Turning should cease during the last 24 hours, which are technically in the hatching period.

Candling

This is a method by which the embryo developing inside the egg can be checked. Traditionally, a candle would have been used rather than the more high-tech gadgetry used today. Nowadays, a strong electric light source is used to illuminate the contents of the egg. An egg is "candled" by holding it in front of the light source. It is easy to candle eggs from an incubator using a modestly priced candling device. Eggs being incubated by a broody hen are less accessible, because the hen will only leave her eggs once a day to look after her own needs. Attempting to take eggs from under a broody hen will cause upset to the hen and is best avoided.

By day five of the incubation period, a good lamp will illuminate the beginnings of the vascular system within a fertile, white-shelled egg. By day six, even with a less powerful lamp, it will be possible to see changes happening within a fertile white or tinted egg. Examination between days seven and ten will reveal the most noticeable feature of a fertile egg – a clear, definite line between the air sac and the rest of the egg. When this is sharp and distinct the egg almost certainly contains a live embryo. Next to this is a spidery form, a few veins near the line between the air sac and the rest of the egg. It may be possible to observe this

spider-like form moving, a sure sign of a live embryo. An infertile egg will remain clear.

Sometimes there will be a less distinct boundary line accompanied by a dark blob that only moves as the egg is turned. In this case, whichever way you turn the egg, the blob moves in the opposite direction. This is usually a sign of a broken yolk. Those new to egg testing may wish to leave these a little longer to see if they develop. However, a broken yolk usually represents a failed egg that has the potential to spread spores and

▼ *From top to bottom and left to right: a clear, infertile egg at 7 days, a fertile egg at 7 days, an egg with broken yolk at 7 days, an egg containing excess moisture at 14 days, an egg with insufficient moisture at 14 days; the air sac content of the egg as it should appear at each annotated day.*

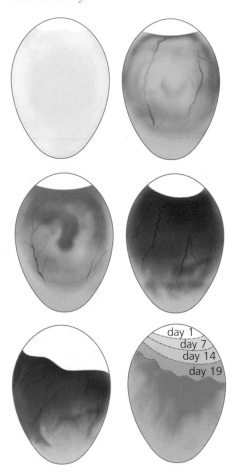

day 1
day 7
day 14
day 19

disease in the incubator. It should be removed as early as possible.

Viewing through the plastic dome of a modern incubator allows adjustments to be made to moisture levels. Some poultry keepers rely on weighing eggs at set stages of incubation to monitor moisture loss, but this takes little account of variations in incubation moisture levels. Eggs that have been stored for more than a week before putting in the incubator will have lost some moisture. Really fresh, disease-free eggs remain the most important prerequisite of a healthy hatch.

Hatching

A hatching unit is only necessary if you hatch enough chicks to justify the purchase, otherwise the eggs can remain in the incubator. A hatching unit is a warm, still air environment like an incubator. The hatching dust, debris and waste from newly hatched chicks can contain diseases that have incubated along with the eggs. Using a hatcher ensures that hatching chicks can be isolated from eggs still being incubated. This makes cleaning the incubator easier, and helps to control egg-borne diseases.

The freshness of the eggs placed in the incubator, and the identity of the breed that laid them, determine the optimum time to transfer eggs from incubator to hatcher. There is an obvious advantage in incubating eggs laid by similar breeds at the same time. Hatching can occur between days 18 and 21. During this stage the eggs do not need to be turned, and should ideally be moved to a purpose-made hatching unit. If the incubator manufacturer's instructions are followed, any hatching problems are likely to be caused by egg storage or by health problems in the parents. Most chicks are dry and strong enough to be removed to a brooder unit 24

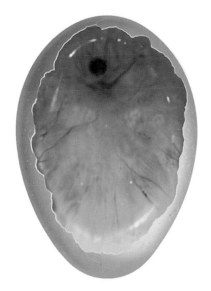

▲ *Day one embryo growth inside the egg.*

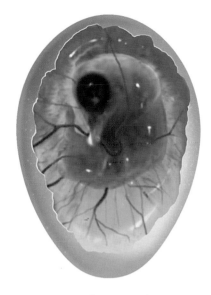

▲ *Day seven embryo growth.*

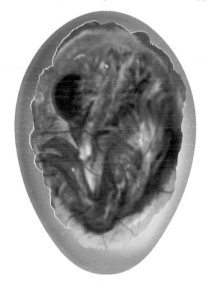

▲ *Day 14 embryo growth.*

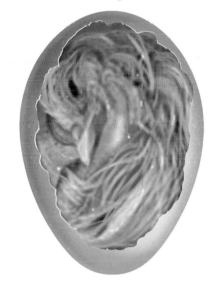

▲ *Final day prior to hatching.*

▲ *Only by candling eggs regularly will you know what to expect inside the egg.*

hours after hatching. During their time in the hatcher unit, hatched chicks live off the egg yolk and it is not necessary to provide them with food. Commercial operators often clear unhatched eggs from the hatchers on the 22nd day, but hobby breeders may often wait longer in the hope of rescuing a last precious emerging life.

Traditionally, the point of hatching was the time when commercial operators boosted the moisture content of the incubator in order to increase chick size. As dry air can toughen the shell membrane, making it harder for

the chicks to hatch, the hatcher should be opened as little as possible during this period. In order to hatch, the chick requires freedom to move within the egg. This means that the air space has to be large enough and there has to be sufficient moisture to prevent the egg membranes from drying out. If moisture has been properly controlled during incubation, difficult hatches are unlikely. The temptation to give chicks a helping hand should be avoided.

The brooding unit

Day-old chicks have a natural resilience, and, given the right sort of warmth and food, are surprisingly easy to look after. As they grow, their housing and management should be tailored to reflect breed feather growth and for a wide range of external conditions.

For days 1 to 28, chicks can be contained inside a brooder unit. This is simply an indoor space warmed by the heat of a reflective infrared lamp. Infrared lamps vary in output from 100–300 watts. Most other lamps do not heat space effectively. Glass reflector lamps break easily; ceramic versions emit dull lighting; while clay-based lamps are extremely durable. The reflected directional heat from the lamps can cause draughts at floor/chick level, so it is necessary to surround the area holding the chicks with a draught-proof hardboard barrier forming an area that can expand in size as the chicks grow. Surrounding the chicks ensures that the entire brooding area can be heated and the heat lamp kept at a reasonable temperature.

Lamp height and resulting temperature are best judged by observing chick behaviour. If the lamp height is correct, the chicks, when not eating and drinking, will be dispersed across the floor area. If the lamp is positioned too high, so that not enough heat reaches the floor area, the chicks will huddle together to keep warm. At this point it may be necessary to deploy an extra lamp. If the lamp is too low, making the area beneath it too hot, the chicks will not sleep directly under it and will constantly reposition themselves away from the heat source. However, as the infrared rays may damage the vitamin content of the chick feed, feeding troughs should be placed outside of their immediate beam.

▲ *Chicks are moved from the hatcher to the brooder, where they will stay until they are sufficiently feathered.*

While a shallow cardboard lid may suffice for feed for the first few days, a quickly growing family will require larger feed troughs. Disposable, but clean cardboard, covered by a shallow layer of clean white wood shavings, provide perfect insulation from a cold, solid floor.

▶ *Clean water founts need to be within easy reach but feed and feeders should be positioned outside of the direct beam of an infrared lamp.*

The growing chick

1 At 30 minutes old, this little chick is still wet, but will soon be a ball of fluffy chick down.

2 At only 5 days old, the first true feathers appear on a young female from a heavy breed.

3 At 7 days old, a Sussex chick like this one will begin to look like a young table rooster.

4 At 9 days old, the chick will be relaxed enough in its surroundings to sit and take a rest.

5 At 12 days old, this chick is beginning to show the typical heavy breed bone development.

6 At 13 days old, the chick's wing feathers are showing definite signs of strong development.

7 At 16 days old, this young female Sussex chick shows significant feather development.

8 At 19 days old, this Sussex shows untidy feather development often associated with males of the breed.

9 At 29 days old the same male is beginning to look like a smart young man.

EXHIBITING POULTRY

Regional and national poultry shows provide an opportunity to see practically every breed of poultry available under one roof. Among the cacophony of noise, smell of ammonia, excitement and general hubbub will be some of the best examples of breed type that you are ever likely to encounter. Showing poultry is a serious business, with exhibitors travelling considerable distances to a national or regional show with their prize fowl for the chance to win a coveted rosette. This is an old-fashioned sport, in which prize money is minimal and the competition is all about the skill of producing a bird that fulfils the exacting criteria of the written standard. For those exhibiting, such events provide the chance to show your best birds and make your mark as a breeder of quality poultry.

Large-scale annual poultry shows are not just for exhibitors, however; those with a specific interest in a particular breed will find most breed clubs take a stand at such exhibitions, providing the perfect venue to discuss the merits and finer points of your chosen breed with the experts.

▲ *Eggs are exhibited in different classes at poultry shows, and judged according to set standards.*

◄ *Regional shows offer the opportunity to exhibit poultry and other livestock, and are a great way to introduce breeds to the public.*

THE BREED STANDARD

The concept of pure-bred fowl being developed to a recognizable standard dates back to the craze for poultry which spread throughout Britain and America in the mid-19th century. At this time, fowl from Asia arrived on the poultry scene with much fanfare and media hysteria.

The first national exhibitions of poultry began in the 1850s. These were events where breeders from far and wide brought the best examples of their breeds to be shown and compared with others of the same breed, as well as with other breeds. It was the exhibition scene, rather than the poultry industry (still in its infancy at this time) that determined what characteristics constituted a breed profile. With new birds from foreign lands being shown, poultry judges began to propose and compile written breed standards for the newly imported fowl, as well as for the local types for which appearance and attributes were well known and understood. The standards determined the accepted colours, weight, dimensions, character and visual appearance to which each breed must aspire to be considered a pure breed of distinction.

Prior to this date, poultry breeds were developed for their utility qualities. The best birds were either the most productive, or produced a heavy carcass with visually appealing white meat that was saleable at a high price. Each breeder would have been aware of the requirements for the breed, and would have endeavoured to reach that standard with their breeding fowl. At this time, the standard was an acknowledged set of requirements rather than specific written details. Soon the Cochins and Brahmas, which were Asiatic imports, were being bred to meet the written standard.

Written standards can be complicated. For instance, the visual characteristics must take into account variations in comb shape or feather type. In some breeds, such as the Dorking, for example, it is possible to find complex rules that permit some

colours to be bred and shown with single combs, and others to be shown with rose combs, or both.

It is likely that in the need to be specific, the standards may have over-emphasized the size and fluffy feathering of these Asiatic breeds to the detriment of their utility potential – a trend that persists in many breeds to this day. Later breeds like the Rhode Island Red and Sussex were bred and judged to standards that put the utility considerations of the breed first.

These standards are accepted in each country by an accredited body. In Britain this is the Poultry Club. Similar clubs and societies exist in other countries worldwide to set the standard for acceptable colours and characteristics for each breed. There are differences between countries in the standards for each breed. Colours and even breeds that are known and

▼ *The Silver-laced Wyandotte was originally selected on the basis of the attractive and ornamental lacing seen on this example. Such lacing is difficult to perfect and is a challenge for breeders.*

▼ *The White Wyandotte broke many early laying records, but as exhibition strains of Whites were selected for excessive feather and fluff, these strains became poor layers.*

▼ *The large fowl that arrived from Northern China in the 1850s introduced many new traits to breeding stock, including an excess of foot feather and fluff as in the Cochin breed.*

▲ *The Suffolk Chequers bantam has yet to be accepted as a standardized breed.*

recognized in some countries are not accepted in others. It may often take years for a breed to be accepted.

Early standards

Though written standards were officially logged from the mid-19th century, breeders and enthusiasts prior to that date would have had agreed acceptable standards for their breeds. Much of the indigenous poultry of Northern Britain, Northern and Central Europe, Turkey and the Eastern Mediterranean countries had distinct feather and colour patterns which had long been selected to conform to very precise local "standard" patterns. Nearly all of these consisted of combinations of black spangling or pencil markings on a gold or silver ground colour. Bolton Greys, Manchesters, Moss Pheasants, Yorkshire Hornets and Mooneyes are regional breeds that would have been exhibited locally long before the first organized shows. Today there are more breeds than ever, and with new breeds being created and imported, the trend looks set to continue. All need to be exhibited to an agreed written standard.

▲ *More than 200 years ago, Sebright breeders found that they could get perfectly laced male tail feathers by selectively breeding for female-type feathers: 70 years later this characteristic became part of the breed standard.*

Manmade breeds

The first poultry shows changed the whole perception of poultry. New breeds were made and selected to meet specific requirements of markets and commerce. Later, when the value

▼ *When Langshans were first imported from Northern China they would have looked like the earliest Cochins.*

of selective breeding was fully understood, commercial strains of many breeds were developed that became the basis of modern hybrids.

Most of the old standard breeds have, for a time at least, played their part in economic poultry development. Selection on the basis of features such as excessive fluff and feather has diminished the usefulness of many exhibition strains of previously important utility breeds. However, it is largely thanks to poultry exhibiting that most of the breeds of the last 150 years are still with us. Breed standards remain the means used to define a breed.

While many of the original standards were written with the breeds' utility in mind, it is now more than 40 years since standard pure breeds played a major role in either egg or poultry meat production. As a result, commercial interests no longer influence the way that breed standards have been interpreted. Those flocks of pure breeds that once supplied the industry with breeding fowl to produce meat or eggs in quantity have been almost entirely replaced by hybrid-breeder flocks.

▼ *The Cochin group of fowl introduced Buff coloration to the western world.*

THE POULTRY SHOW

Enter the world of the poultry show and you will find yourself in a large hall filled with row upon row of wire cages containing every breed of poultry imaginable. Exhibition birds are organized according to similar criteria throughout the world.

Show criteria are the same whether you are visiting a small-scale local show or a huge national exhibition. Exhibits of each breed are positioned next to each other so that like can be compared with like. Large and bantam versions of the same breed are judged separately, and may be located in different places within the exhibition hall. All breeds that belong to the same class appear in close proximity to each other so that the best of each breed within a class can be compared to other winners. Visit any of the large poultry shows and you are sure to find examples of local or rare breeds.

Poultry are divided by class, which is defined by the world geographical region responsible for selection and development of the breed. Sometimes this can be confusing, for example,

the Hamburg breed hails from the Netherlands but is classified as English, because English breeders were responsible for developing the breed that exists today. The confusion arises because the breed's name suggests it should be of German origin, but it is, in fact, named after the port through which it passed on its way to England. The classes that exist are Asian, American,

◀ *Poultry are exhibited locally at agricultural shows during the summer months. Local poultry clubs often take an interest in such shows, and this can be a good opportunity to locate breeders and even buy birds.*

Mediterranean, European, English and Rare, the latter being the only class not defined by region. The occasional appearance of breeds that are not well represented in their homeland has become a regular feature of rare breed classes at major shows; rare breed categories include minority breeds. The same categories exist in the bantam classes. Turkeys and waterfowl are also judged according to breed. Some breeds and varieties that are recognized in some countries are not acknowledged in others.

▼ *Trained show entrants know how to display themselves to advantage. Many are happy to stand to attention and move to the pen front when being viewed.*

▼ *Eggs that are exhibited at poultry shows are often of interest to the public. Judging the contents class means looking for the same qualities as the discerning cook.*

EXHIBITING POULTRY

While only a small percentage of people who keep and breed pure-bred poultry are likely to become regular exhibitors, a visit to a well-run show can add much to our understanding and appreciation of those breeds and the standards to which they are expected to adhere.

With thousands of entries, the great national shows are a wonderful window into the world of poultry breeds and the numbers of breeds and varieties that are available.

For the competitively minded, the opportunity to exhibit prized fowl and collect a winning rosette or trophy on the show circuit may be the most important reason to keep poultry. Others will find that showing poultry and attending shows is enjoyable and adds much to an interesting hobby. Poultry exhibiting is likely to attract different people for very different reasons and there are plenty of other reasons to exhibit and attend poultry shows.

For those interested in the conservation of rare breeds, poultry shows are essential. Without the persistence of a few breeders, many

▼ *The best of each class is removed to winner's row to be judged against one another. The supreme champion for each show will be decided here.*

▲ *Shows are the best place to see good examples of each breed type. The trained eye will be able to detect differences between exhibits of the same breed.*

breeds, particularly those with less utility merit and therefore less obvious reasons for keeping them, would have ceased to exist. The fact that these breeders have made the effort to gain recognition for their

breeds by showing them in rare breed classes means that the poultry world has a richer heritage to draw on in the future. The exhibiting of good examples of specific breeds helps to generate interest from new breeders

▼ *Birds that are used to being both confined and handled will suffer less stress than those taken direct from a poultry run to a crowded show.*

interested in continuing to preserve this heritage. Endangered regional breeds such as England's Derbyshire Redcap may be shown locally in order to maintain local interest for strains genetically adapted to a specific environment. Conservation of these important old breeds is in the hands of a few small-scale breeders, hobbyists and the specialized societies that support them, their exhibition classes presenting the results of centuries of careful selection.

Breeders who keep birds with specific or unusual colour markings, patterns or lacing that are difficult to perfect may only breed a near-perfect example every few years. As a result, the chance of seeing perfect examples of such breed coloration are few and far between. A breeder may exhibit a perfect specimen mainly so that he or she can show it to his or her peers, if only to discuss how it compares with previously shown breed examples.

▼ Having first gained an overall impression of a bird's shape and type, a judge goes on to examine it critically in greater detail.

▲ Staging any show can involve a lot of paperwork. Making sure that every entry receives the correct prize card and award often relies on the efforts of a small group of volunteers.

Beginners unfamiliar with the visual appearance of different breeds may find that few of the exhibits are labelled, though most classes are. This can be frustrating, but show genuine interest in any exhibit, or ask about any breed, and you are likely to find yourself with your own show guide or an introduction to a local breeder.

Local shows

With a more relaxed and slower pace than national shows, local shows may provide a better introduction to poultry exhibiting, as well as to local exhibitors and poultry keepers who are in the best position to help the beginner. Would-be exhibitors can see how others carry birds to shows and put them in pens.

Due to the internet and an increasing number of quality poultry magazines, it is easy to find details of breeders and dealers, but there is nowhere better than the local poultry

show to find those breeders with the best breeds in the area. Local breeders may have surplus breeding stock for sale, or in some cases may be willing to sell reliable hatching eggs from healthy stock that you can incubate and hatch if you have the facilities. Such breeders may also accept requests to breed a few extra birds that they would be willing to sell. Above all, the local show is the place to see the best examples of poultry breeds that do well locally. It is a chance to make contact with like-minded people, catch up with old friends and support the efforts of those who have helped to organize the show.

There will be fine examples of many different breeds on display. However, the breeds that are kept locally will obviously be greatest in number. This may be because a good breeder has successfully sold his or her birds into the local community, or in areas that experience extreme temperatures, it may be because certain breeds survive better in the environment than others.

New or would-be exhibitors who have bought stock from a breeder who has the long term interests of that breed in mind will be actively encouraged to enter a local or regional show. They are likely to have the benefit of the original breeder's expertise to guide them through the preparatory steps necessary to exhibit their fowl competitively. Over a cup of tea or coffee, important local contacts can be made at the show, and it should also be possible to learn about other activities such as open days, poultry courses and discussion groups that the local club organizes. It is how well a club looks after its newcomers as well as existing members that boosts and retains membership, and encourages new members to take up exhibiting.

As most shows allow "open" judging, meaning the public can view the judging process, it is possible to get some idea of the finer points of exhibition judging. You may see a judge or steward looking for broken or missing wing feathers, incorrect skin coloration and even the discovery of a poultry flea or mite.

While those with a specialist knowledge of a single breed may be asked to judge that breed, all experts engaged to judge general classes will have passed practical and written tests on a group of breeds.

By mid-afternoon, most of the prize cards for the best of breed will be in place. The best breeds in each class are compared and the best class champions are moved to their place in the champions' row. It is from these that the best bird in the show will be decided. Every show will have unexpected results — perhaps a juvenile beating the experienced exhibitor, or a first-time exhibitor winning a major award.

Agricultural shows

Poultry are often shown at agricultural shows, which are held outdoors in the summer months. Unfortunately, this coincides with that period of time when most adult fowl are losing their feathers, beginning to moult or showing signs of breeding-pen activity, so one cannot expect to find many good examples on show.

Judging

As most judging at local shows does not commence until mid-morning, an early visit to the show hall provides a good opportunity to watch the hustle and bustle of "penning", and the hurried cosmetic touches that are made to the poultry. Old hands may well help juveniles or absolute beginners prepare their birds for exhibition.

Watching the judges and stewards get the exhibits out of their pens is not only a good way to learn how to handle fowl, but also provides those considering exhibiting poultry with an idea of the amount of handling their birds need in order to be calm when held by unfamiliar hands. Teaching the birds to stand and remain quiet is essential, and many

▲ *New judges can learn much while deliberating with more experienced judges*

exhibitors have plenty of tricks for training their birds. Experienced exhibitors may select their birds for exhibition from a young age, and start to train them as soon as they show promise of fulfilling the criteria for the written standard. Others will leave their poultry to grow up as part of a small flock in a healthy outdoor environment, and then select the best to train for the show pen.

Whatever the nature of the breed, the more excitable types must be accustomed to being handled and confined prior to showing. At exhibition, birds are placed in small cages, in which they may have to stay for up to 36 hours. Some breeders place their birds in training pens, visiting to give a tidbit of food, and then when the bird is totally relaxed, they will gently pick it up a few times and replace it in the pen in the same way that the judge will handle it. At exhibition, thus familiarized with being handled, birds will come to the pen front to see if the hand that picks it up has food to offer.

▼ *A knowledgeable judge will be able to tell much about the overall condition of a bird in his hand.*

PREPARING FOR THE SHOW

Weeks before the show, preparation begins. Owners train their birds to cope with the restricted show pens and stand to attention. Ultimately, the bird has to present the best characteristics of the breed and measure up as closely as possible to the written standard.

The first requirements of a show bird are fitness and health. If a bird is not in the best of health, then there is no point in showing it. A whole range of specialist poultry feed is available that may help to condition show birds.

Some dedicated exhibitors start work on the showing posture of their birds as chicks. Birds need to be used to being handled, and at ease with being confined to a small show pen for the duration of the show. A show training room set up at home, composed of pens not much larger than show pens, will help accustom young birds to being confined before being exhibited at their first show.

Breeders of true bantams may generally keep their tiny charges in pens not much larger than those used at exhibition. These fowl should be accustomed to being handled by a dedicated breeder, who devotes time and attention to them.

Breeds like the Orpington, which are docile, require little show-pen training, but they do need to be sufficiently alert and attentive when viewed to come to the pen front. Indeed, some of the heavier breeds, such as the large Buff Orpington, can be almost too immobile to be really attractive. Judges of active breeds such as the Egyptian Fayoumi accept that agility is a breed type, but the birds still need to be accustomed to being handled. A little hard-boiled egg, mixed with chick feed and offered to the bird from the hand, will get it used to being in close contact with humans. With the expectation of a food treat, small birds may walk straight on to the judge's hand as the show pen is opened.

Washing poultry

To be seen at their best, most soft feather poultry will benefit from being washed with specially formulated poultry shampoo prior to being exhibited. White varieties and those breeds with very fluffy feathers will certainly need washing before a show. It was once thought that cleaning black fowl "washed out" their lustre and natural beetle-green top colour, but this is no longer believed to be the case.

Washing and drying poultry

1 Use a small bath or sink filled with warm water for bathing. If the water feels warm to your bare elbow, then the temperature will be acceptable to the bird. The bird should be carefully but firmly submerged, leaving just its head out of the water.

2 Holding the bird firmly, gently pour water over it to wet it thoroughly. Keep the water tepid and keep hold of the bird. Birds unused to such treatment may try to break free. For this reason, a small bath is a good choice, leaving no room for flapping.

3 Use a damp, soft, small sponge to gently cleanse the comb and wattles.

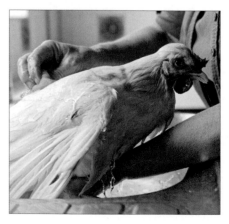

4 Add a squirt of shampoo to the bird's back.

5 Thoroughly massage the shampoo into the fluffy feathers and carefully rub it into the wing feathers using a small sponge. Rinse thoroughly with tepid water.

6 Wrap the bird in a warm towel to remove excess water. Wet poultry feathers can hold a lot of water. Pat gently dry, but do not rub or you may damage the feather structure.

7 Use paper towels to rub down the bird and remove the excess moisture before blow-drying. Pat dry in the direction of the feathers.

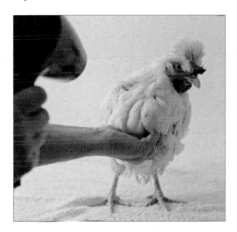

▼ *White-feathered birds will almost certainly need to be washed at the last minute to ensure that they are in pristine condition when they are shown at exhibition.*

8 Blow dry the bird using a medium heat setting on a hair dryer, starting with the area around the heart and lungs. Check the bird is not too hot.

9 A bird that is used to being handled will move to accommodate drying.

Clipping the beak

1 The tips of beaks, like toenails, are usually kept in trim by running about and pecking feed from hard surfaces. Birds kept in a protected pre-exhibition environment may need the upper mandible to be trimmed.

2 Trim away any horny toenail-type material carefully with nail clippers. The bird will feel no pain, but like many other simple poultry procedures, this is best done by an experienced poultry keeper.

However, specialist shampoos are now available to enhance nearly every feather colour. These shampoos can be purchased at all major poultry shows, as well as from some feed barns. Specially formulated shampoos are kinder to the feather structure of the bird than washing-up liquid or

standard hair shampoo. The feather web has interlocking characteristics and so most feathers retain their normal structure.

Since showing poultry is a competitive business, many fanciers develop their own formula for shampoo, which is tailored to their particular breed's show requirements. Most poultry clubs hold demonstrations on how to wash and prepare birds for show. Drying the fowl properly is as important as washing it, and will be easier if the whole drying room can be kept at a minimum 20°C. It is at this point, provided the bird is tame and accustomed to handling, that most fowl appear to enjoy the washing and drying process. By the time the fluffy areas between the legs and under the tail are dry, many birds will start changing position to dry those areas that are still wet. Some exhibitors

leave their fowl to finish drying in a warm room once the initial wetness has been removed from the feathers. Others wrap their bantams in kitchen paper for up to 15 minutes, and then leave them to slowly dry and "fluff up" in a warm room.

Final preparation

Birds kept in conditions other than free range will occasionally require their toenails trimming and may suffer from an overgrown beak. These are both regarded as serious exhibition faults if shown on an exhibition bird. Dry and dead skin on combs, face and wattles can be carefully removed with a toothbrush or soft nail brush a day or two before the show. The final preparations can be left until the morning of the show. Most of this last-minute preparation involves ensuring that mouth, face and comb are scrupulously clean. You may see exhibitors using "secret" lotions to brighten red areas of the face, or wiping baby oil off the kid glove-textured lobes and applying a dusting of fine powder. Some will simply use a smear of petroleum jelly on the face feathers, other will use different products with a slightly astringent quality. Whatever is used should be washed off carefully after the show. In fact, it is these last-minute preparations that add to the excitement and drama of show day for many exhibitors.

▼ *"Closed" leg rings can only be bought from national poultry clubs and must be fitted while the bird is young and small enough for the ring to be slipped over its closed foot. The ring indicates the original breeder and the year that it was hatched. This can deter thieves from swapping birds in show pens.*

▶ *Exhibition birds that are used to being handled and preened for exhibition will become familiar with the "unnatural" environment of washing and drying.*

White birds that are inclined to look tan or yellow, and buff birds that look bleached, will have to be shaded from too much direct sunlight. The large white lobes of breeds like Spanish and Minorca may become chapped or blistered by cold winds when in the open. Another form of blistering may occur when males are being prepared for show and confined to smaller pens. As this has long been thought to be linked to overeating and a change in energy levels, earlier exhibitors used to add a measure of bran to the pre-show diet. Other than

a quick dust with baby powder to dry them, there is no last-minute fix for these blisters, which were probably caused by knocks.

Leg mites should be treated as a matter of routine, but even birds kept in reasonably clean conditions will have some dirt that has found its way under the edges of the leg scales. To remove this, exhibitors first soak the legs and then carefully clean under each scale with a cocktail stick. Others have found this dirt easier to dislodge after a few hours' soaking in industrial hand cleaner. Combs and

wattles that have become dirty and discoloured will benefit from a thin coating of petroleum jelly that should then be wiped clean. Most fanciers will have their own secret dressing recipe.

Having bred a good bird and then spent days or weeks on pre-show preparation, every exhibitor will want to give every one of their exhibits the final show touches. For many exhibitors, the pre-show preparation is often a means of relieving their own pre-show tensions. However, many well-prepared exhibits are spoilt by evidence of the bird's breakfast.

Final show touches

1 Cleaning and putting a spot of comb dressing on a flat single comb is relatively easy, but reaching every nook of the intricate Sebright comb will take time, patience and the assistance of a cotton bud.

2 A final wipe of the beak and wattles makes a pristine impression. It is surprising how often a judge will find an otherwise beautifully prepared exhibit spoilt by the remains of the bird's hurried breakfast.

3 Nearly every exhibitor will have a favourite comb preparation, varying from petroleum jelly to baby oil, with some containing just the right proportion of astringent. Wash the preparation off after the show.

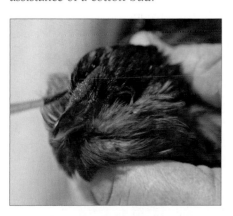

4 Dust the bird's muffs and beard with an artist's paintbrush to ensure they are in pristine condition.

5 Scales on the legs and feet may benefit from a rub with baby oil. Remove dirt at the same time.

6 A quick check for dirt under the toenails is an essential part of last-minute show preparation.

EXHIBITING EGGS

Eggs are often shown at the same shows as poultry and the classes are every bit as competitive. Viewing the trays of specimens on show is a good opportunity to see the assortment of differently coloured eggs as well as the varying sizes that different poultry breeds lay.

Not everyone who is a member of a local poultry club will have bred the poultry they own, and for that reason they may see little point in exhibiting their prized pets, leaving that opportunity for the serious breeders. For some owners, the task of washing and preening their poultry to exhibition standard fails to hold appeal. For those interested in the utility aspects of poultry keeping, however, the opportunity to show and compare quality eggs may have appeal.

Egg classification will usually include single eggs, threes, and less commonly six or even a dozen eggs. For the classes of multiples, all the eggs shown are expected to be the same size, shape, texture and appearance so these classes provide a great challenge. Often exhibited at the same show as poultry, a long trestle table containing plates of eggs are judged with equal seriousness to the birds. Washing is not allowed.

There are exacting rules that define the appearance of a winning egg.

▲ *Eggs are judged on their contents as well as exterior appearance.*

Bantams, pullets and large breeds all lay slightly differently shaped eggs; however, judges accommodate this and the weight of the egg has to be appropriate for the breed.

Egg judges and exhibitors are enthusiastic, and many are willing to share their knowledge to help beginners. Unusually, the mid-brown or tinted eggs laid by hybrid hens can compete with pure-breed eggs.

▲ *The "three brown eggs" class includes all shades from light to mid-dark brown and can include hybrid eggs.*

▲ *The "dark brown egg" class results in strong competition between owners of the Welsummer and Marans breeds.*

▼ *Bantam and miniature fowl weights have increased, but the standards for their egg weights have remained constant.*

▲ *The class for "eggs of three distinct colours" requires finding similarly shaped and sized eggs laid by different breeds.*

▲ *At a few days old, white eggs may display tell-tale grey spots, so the "white egg class" is not always well supported.*

▲ *Judges may sometimes move a plate of eggs to a more appropriate colour class.*

Egg colours

Dark brown eggs were partly responsible for the 1850s poultry craze, since they had never been seen before. A really dark brown shell has remained as the mark of exterior egg quality. The public assumption that brown eggs are healthier than white is completely unfounded, however.

▼ *Unlike the bantam egg weights, the standard weight for large fowl eggs takes a more logical approach. Size is not necessarily a deciding factor, but should be appropriate to the breed and species.*

THE PERFECT EGG

In external appearance, the prize egg has to have good shape, which means that it has an even dome at the widest end, or, as one leading egg judge defined it: "one that sat perfectly in an egg cup". Eggs should be longer than they are wide, and the top should be broader than the bottom. The whole should be curved, and taper to a narrow bottom, which should not be pointed.

The shell texture has to be smooth, without lines, bulges, blemishes and rough edges. The shells should be clean, though unwashed, and should be freshly laid. Stained shells are not permitted. Eggs can be shiny or matt; the colour has to be even. Speckling and mottling should be even. Egg colours can be white, cream, brown, mottled or speckled, olive, blue, green and plum.

Inside the egg, the air sac should be visible, with the membrane attached to the shell. The air sac should be to one side of the domed end of the egg. The yolk has to be uniform, bright, golden yellow, and unblemished. It should be rounded, standing proud of the albumen, and there should be no sign of embryo development. The albumen has to be clear and blemish-free, with two distinct layers; the thicker albumen closer to the yolk must be distinct from the thinner albumen.

Stale eggs are disqualified. Those with a bloom that suggests absolute freshness are well-placed. The judge will gently shake an egg that he suspects of being stale and will often tap a shell with his fingernail and listen for a tinny ring that indicates a sound shell

Even in today's competitive egg classes it is the darkest eggs that attract the attention. While one perfect egg in the single egg class can and often does win the "best egg or eggs" award, three perfectly matched eggs will score higher than a single exhibit. Similarly, six identically coloured, perfectly matched eggs of any colour will always be high on the priority list for the award of "the best eggs in show".

Newer hybrids are now being marketed on the basis of their dark eggs, though few yet lay the quality egg of long-established strains such as the Welsummer and Marans. White eggs must be perfectly clean, but as their nearly translucent shells are inclined to quickly show grey "age" spots, they are among the hardest to exhibit to perfection. Blue eggs that are still seen as a novelty are not always full-sized. Olive eggs, or those with a green tinge, are produced by fowl created by crosses between blue and very

dark egg laying fowl. Their appearance at shows can enliven mixed or novelty egg classes.

▼ *The egg classes always interest visitors to summer agricultural or poultry shows. Often there is a class for children for the best decorated egg.*

KEEPING OTHER DOMESTIC FOWL

This chapter includes domesticated ducks, guinea fowl, geese and turkeys – birds bred to provide meat and eggs for human consumption. These birds are traditionally associated with a farmyard or backyard setting, more so than hens. Many of these birds are large and require plenty of space and, for ducks and geese, a body of water.

Often poultry keepers begin with hens and graduate to keeping other types of poultry or domestic waterfowl. Few poultry keepers have the space to allow different species to integrate, though it is possible for those who work on a small scale.

In terms of welfare, housing, health and economic considerations, separate management regimes for each poultry type are generally the norm, particularly on farms, which have to adhere to rigorous guidelines regarding health and welfare. Many breeds of ducks and geese can make fabulous pets, and if you have the space and time available to look after them, they can be very rewarding to keep. Being larger birds, turkeys are most likely to be reared for a special occasion feast than as a pet.

▲ *Swiss Crollwitzer turkeys are more strikingly marked than the older Pied variety that they have largely superseded.*

◄ *Ducks kept on farms and holdings have always been kept for eggs or meat.*

KEEPING DUCKS

Most domestic ducks are larger than their wild counterparts, and are the result of selective breeding. Like their ancestors, they are mostly aquatic birds. The adult female is known as a duck and the adult male as a drake.

Nearly all the pure breeds of duck were selected, bred and developed from the end of the 19th century for either meat or egg production. Today, people keep ducks for the same reasons that they keep chickens, for meat, eggs and also to exhibit, as well as for purely ornamental purposes. Lighter breeds would historically have laid more eggs than the heavier breeds, which were generally kept for their table properties.

Duck meat

Most of the world's commercial duck meat is now produced by ducks descended from Aylesbury and Pekin bloodlines. Aylesbury and Pekin ducks and the Aylesbury-Pekin cross have been developed to make good table birds. Aylesbury have pure white flesh and feet, a feature that has been

▲ *Duck eggs can be much larger than hens' eggs, and are highly prized by cooks.*

highly prized in the Western world, since it was thought to be the sign of high-quality meat. Pekin have yellow skin, legs and beaks, and their commercial crosses also have yellow legs, beak and skin. However, Rouen and Muscovy ducks, which have dark, richer-tasting flesh, are also bred for meat. The Rouen is slow to grow and mature from the duckling stage. The slow-growing Muscovy can

be crossed with other heavy, fast-growing table breeds to produce an infertile hybrid generation that have huge, superbly flavoured carcasses. Pekin ducks are often intensely fattened and at 14 weeks they are slaughtered in the industry for meat. Keeping ducks for meat beyond this point does not make commercial economic sense, since the meat will not improve in taste and the bird will continue to cost more in feed.

People who keep birds for meat either keep a few birds for themselves, or may run small-scale businesses with a relatively fast turnover of the product. Birds reared for meat can be bred from existing

▼ *Muscovy and Pekin crosses are often reared in close confinement. These free-range ducks are enjoying their daily dip.*

DUCK DOWN

Soft duck down is one of the best insulating materials, and is generally used in duvets and clothing. It has a high thermal content. Warm air becomes trapped between the fibres helping to provide heat and insulation. Down feathers are those closest to the bird's body, generally beneath the outer layer of plumage. The best-known source of down is from the wild Eider duck.

▲ *Ducks will cause considerably less damage to your garden than hens.*

stock If you have the facilities to hatch ducklings, eggs can be purchased from reputable suppliers, or they can be bought as day old chicks from hatcheries. Some pure breeds (such as the female lines of modern Aylesbury-Pekin crosses) that used to be kept for meat production have been selected to lay 200 or more eggs in the year-long breeding season.

Duck eggs

Breeds such as the Indian Runner and the Khaki Campbell are the most productive egg-layers. If you want eggs, then these are the breeds to choose. Utility strains of Indian Runner ducks can lay 250 eggs per year, while commerical Khaki Campbell ducks can lay up to 300 eggs in the same timeframe. No other pure breed of duck can convert food into eggs at the same production rate as the Khaki Campbell.

The lighter utility strains of ducks may come into lay at 18 weeks and may, with careful management, continue to lay well for up to two years. At this point, they no longer convert feed to eggs at an economic ratio. Some breeders retain older ducks for breeding, and these may live on for many years as family pets.

Duck eggs are usually larger than hen eggs, weighing about 75g/3oz or more. Shell colour can vary from shades of pale green-blue to white. They have a slightly higher fat

▼ *Few strains of Khaki Campbell lay as well as those previously kept on commercial duck farms.*

content and oilier texture than hen's eggs and are richer in flavour, which makes them ideal for baking. Duck eggs are often laid in muddy places so they should be cleaned and thoroughly cooked. They are available from speciality food shops.

Ornament and pet

Ducks can live for up to 15 years depending upon breed, making them quite a long-lived pet. Many ducks have striking plumage as well as appealing characters that make them interesting garden companions. In many breeds ducks and drakes have very different feather patterns. Drakes generally have far brighter and more flamboyant plumage than ducks, but lose much of it for a few months each year as they go into a post-breeding summer moult. During this period drakes temporarily revert to the muted shades normally associated with the female members of the family.

If you have a large pond, or live near open water, then keeping pure breeds of ducks can be an appealing and rewarding pastime. Many breeders are happy for their pure breeds to live alongside wild ducks.

CARING FOR DUCKS

Ducks are relatively easy creatures to provide for. As aquatic birds they need a body of water, although how much depends upon the breed of bird. Some require water for exercise, washing and preening, while other breeds will not mate and breed without an expanse of water.

Ducks require a clean, dry shelter, particularly from extreme weather conditions, in which to preen and rest, as well as protection from predators and a ready supply of food.

Buying ducks

Pure duck breeds can be bought as eggs to hatch at home, as day-old ducklings or as young adults. Ducklings require different housing and feed than adults, and may also require heat for the first seven weeks of life. Unless you are able to provide such facilities, buying young adult birds from a reputable breeder may be the best choice for stock, especially for beginners. Buying eggs to hatch is a possibility for those with facilities. Check local press, or contact your local poultry waterfowl club for breeders of the type of bird in which you are interested. Before you buy,

▲ *Ducks may enjoy the security of a nesting place on stilts, but to be foxproof it must be surrounded by water.*

however, ensure that you have the right sort of housing and a sufficient area of water for the type of duck that you are going to keep.

▲ *While most ducks lay eggs first thing in the morning in nest boxes within their overnight houses, separate outdoor nesting places can be provided.*

Housing ducks

The amount of land and water required to keep and breed ducks varies according to breed type. Most ducks spend their time outdoors. Would-be owners with running water or a pond will be able to choose from a wide range of breeds, while others may be restricted in their choice of breed by the amount of land and water that they have available. Breeders of small ducks will be able to keep a family of pet ducks or exhibition Call ducks with as little water as a baby bath on raised decking, in addition to a small area of land on which to exercise. Such a restricted living space, however, will need to be accompanied by the highest welfare standards. Breeders of the large and stately Aylesbury will need to provide a considerable body of water for the birds to mate and breed for any measure of success.

Handling a duck

1 Pet ducks may get used to being handled, and will accept being gathered and held in one hand or arm without fuss.

2 Those unused to handling ducks should pick up the duck by firmly grasping its wings together – an experienced breeder will demonstrate.

Unlike hens which perch at night, ducks sleep and lay their eggs at floor level. Their housing needs to be designed with a door that allows human access to change the bedding and clean the nest boxes at regular intervals. The housing should be located on a raised, well-drained area. The flooring should be covered with straw or hay, rather than wood shavings, which can cause digestive problems if the birds eat it. Ducks lay nearly all of their eggs early in the day. In order to prevent eggs being laid in water, some breeders do not let their birds out of the housing until after 9am.

Feeding ducks

Ducks can be messy feeders. Their drinking and communal bathing habits result in spilled water, and dabbling in the mud may not make them the most appealing housemates for other poultry. Ducks eat grasses, aquatic plants, fish, insects and other small water-dwelling creatures, and their beaks are adapted to pull up food from beneath ground level. Ducks have different nutritional requirements at different ages. Formulated adult feeds that are made of predominantly wheat and soya are available as a supplement to the foraging diet. They contain many other essential minerals and trace elements. When natural feed is plentiful, very little food supplement

Cleaning out a duck pond

1 Call or other small ducks can be kept in a small garden. With care and good management, these pets can live on a small pond or even a baby bath, although this will need cleaning on a regular basis.

2 Algae soon builds up on the surface of artificial ponds. This is best removed with a soft brush before it can build around the edge of the pond or bath.

3 After a rinse, the water can be replaced. As ducks splash water everywhere, some owners may place the bath on decking.

4 Pet ducks will soon learn to interact with their owner, and may start to ask noisily for more water to top up a half filled bath.

is required, but during short, cold, winter days, or when highly productive ducks are in full lay, they need to be given larger quantities of supplements. Ducks kept on ponds and open water may ignore clean drinking water, but those kept in less natural situations should always have a fresh supply.

◀ *Heavy rain turned this duck run into a temporary swimming pool.*

Changing bathing water

Ducks are undoubtedly messy creatures, and any bathing water set down for them will not stay clean for long. Although they prefer a large pond or bath, using a smaller, lightweight bath will ensure that regular changing of the water is a less onerous task and therefore is less likely to be neglected. A pair of ducks will be happy in a bath that is about 70cm/28in in diameter.

BREEDING DUCKS

All breeds of domestic ducks descend from the wild mallard and will readily interbreed with wild birds. Anyone wishing to breed pure-bred ducks that conform to the breed standard may have to keep their birds in separate pens that are covered with nets to prevent crossbreeding with wild males.

Breeders prepared to go to the trouble of breeding their own ducks will have a breed standard or other criteria, such as egg numbers or meat quality, in mind when buying or selecting an adult breeding group.

While one light-breed drake may readily mate with several ducks, very heavy exhibition drakes of breeds such as Rouen and Aylesbury, that find it easier to mate on water, may be able to cope with no more than three or four females. An excess of drakes can make life unpleasant for the ducks that share their space.

Hatching the eggs

Fertile eggs intended for hatching should be collected first thing in the morning, as soon as possible after they are laid, and carefully stored and

▼ *These little Call ducks are now popular as garden pets.*

turned until they can be put under a suitable broody duck or placed in an incubator. While some domestic ducks make good mothers, many small-scale breeders rely on foster mothers, including broody hens or a purpose-bred crossbreed bantam hen willing to sit on the eggs for 28 days (33–35 days in the case of Muscovy ducks). Larger quantites of eggs may be better hatched in an incubator, with the newly hatched ducklings being reared with the aid of an artificial brooder. The advent of very reliable, moisture-controlled small incubators means it is now far easier to hatch ducklings artificially, leaving the ducks occupied with laying eggs.

Sexing the ducklings

Luckily, vent-sexing chicks is easy, and given a little tuition, most breeders will learn how to gently press the vent area to reveal emerging organs.

▲ *Muscovy ducks that are willing to sit for a full 30 days can make satisfactory surrogate mothers for other breeds.*

Young and adolescent female ducks soon develop a characteristic, loud quack, while the drakes only manage a quieter, almost hoarse or hollow, note. Drakes also develop characteristic curled feathers immediately in front of their tails.

Rearing ducklings

Ducklings can be reared by three methods: under the mother duck; under a foster mother, which may be a hen or another duck breed; or in an incubator. Small clutches of ducklings can be raised effectively with a mother duck or mother substitute. Ducklings do not require heat for as long as chicks, and may soon spend less time with their mother. Since ducklings may not need artificial heat provided by the breeder for too long, vigilance is essential. As with chicks, an infra-red heat lamp within a draught-free enclosure allows growing ducklings the option of sitting under the lamp, or moving to the periphery

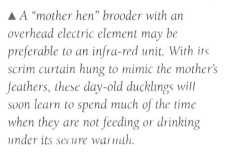

▲ *A "mother hen" brooder with an overhead electric element may be preferable to an infra-red unit. With its scrim curtain hung to mimic the mother's feathers, these day-old ducklings will soon learn to spend much of the time when they are not feeding or drinking under its secure warmth.*

SEXING DUCKS
Hold the duck firmly on its back and very gently press the vent area with your thumb to expose the genitals.

▲ Male ▼ Female

if they become too hot. Carefully observe this behaviour and adjust the lamp height as appropriate to ensure healthy ducklings. Red-tinted or dull heat emitting lamps are available, and are preferred by many breeders who think that the alternative of 24 hours of bright white light creates an unnatural lighting pattern for the young birds. Lamps can be turned off for short periods during daylight hours, when the ducklings spend little or no time under the heat source.

▲ *In this self-coloured variety the drake's curly tail feather helps to differentiate it from the duck, which has flat tail feathers.*

Feeding ducklings
Like adult ducks, ducklings can be messy feeders, and from an early age like to mix food and water into porridge, which they then treat as mud to be rolled in. To combat this, many breeders keep a supply of dry mash or crumbs at some distance from their source of clean water. Some breeders use special duckling feed, but care should be taken to avoid feed containing specialist chick anti-coccidiosis medication. Ducklings being fattened for the table will be fed, first on duckling feed and then on a lower-protein fattening mash.

◄ *Not all broody hens will be willing to sit for the extra week that it takes to hatch ducklings, but are likely to be far more amenable to being moved to a suitable broody coop. Using a hen or an incubator can extend the breeding season beyond the late spring and early summer months when ducks are more often inclined to sit. Ducks will lay and sit in a place of their own choosing, and then bring up a family on their own terms.*

KEEPING GEESE

Like ducks, geese belong to the order Anseriformes. Both species have webbed feet and are developed for an aquatic way of life. They have a strong bill adapted to pecking for food. Male birds are ganders, females are geese and the young are known as goslings.

Geese may have been kept and domesticated by humans long before they kept poultry. Archaeological excavations of Saxon settlements have uncovered bones of geese that had a sufficiently large stature to have rendered flight difficult. As a result, it seems that at this point in their history geese could be considered domesticated, since they would have relied on human shelter to protect them from predators.

Large white geese have been selectively bred, primarily for meat, in northern Europe, including Holland and Germany, for centuries. The Bremen or Embden geese from Germany were notably large. It is likely that strains of domestic geese would have developed in the same

▼ *Once a Christmas goose was commonplace. Popular goose fairs or markets were established in September to sell purpose-bred geese, which were fattened by the buyer ready for Christmas.*

way as other poultry, with different breeders selecting the best in terms of size and shape for meat, or egg-laying capacity, and then using those birds to supply the next generation in the hope that the desired characteristics would be present in the progeny.

Today geese are kept for much the same reasons as other poultry: for meat and eggs; to breed and sell the young or eggs for hatching; for ornamental value; to exhibit at shows; as attractive pets whose antics are entertaining to watch; to keep large swathes of pastureland neatly kept; and, uniquely among fowl, as a deterrent to burglars. Geese are territorial and easily alarmed when faced with the unfamiliar. These are large and strong animals, and will hiss in an aggressive manner if disturbed.

Goose meat – a seasonal treat
European pure-bred stock provided the genetic material needed to create the crossbred, fast-maturing goose

WALKING GEESE TO MARKET
Geese were traditionally walked to market at the end of August or the beginning of September depending upon the time of the harvest. The bird's feet were protected by driving them through a sticky, tar-like solution and then through sawdust before each day's journey. Drovers guided the birds to their destination, which could take several days and cover 30–40 miles. Many were sold along the way. After the harvest, the birds feasted on the stubble in the fields until late autumn, when they were fattened for Christmas.

that was historically commercially bred to supply meat for the Christmas market. The goose lost favour with the introduction of relatively cheaper turkey meat in the early to mid-20th century. Now considered a luxury meat, the goose is increasing once more in popularity.

Unlike other domesticated poultry and animals, the goose has defied the instigators of intensive rearing methods. Geese are not prolific layers and they are one of the few remaining sources of seasonal food. This large,

▲ *Goose eggs are available from specialist butchers and delicatessens.*

fatty bird traditionally survived by pecking about and foraging for any available food. Found on humble farms because it was economical to keep and a good source of meat or fat, a special bird would be fattened for Christmas. It is the traditional celebration bird and has always been seen as a treat: a bird for both the poor and the rich.

Goose fat had all manner of additional uses for country people, including rubbing into the chest to ward off colds, as well as for cooking.

Goose eggs

A gander is not required to produce, only to make fertile, eggs. Weighing in at about 200g/7oz each, goose eggs are twice the size of hens' eggs. The shell is chalky white and very hard. Goose eggs need washing and thorough cooking to kill any harmful bacteria that may be on the shells. In flavour, they are milder than ducks' eggs but stronger than hens' eggs.

Ornament, exhibition and pet

Most breeders are interested in selecting strains of pure-bred geese for utility properties, but others will find that breed characteristics are the appeal of keeping geese. Pure-bred and exhibition geese may be more

▲ *Geese are an effective deterrent to burglars and will hiss and show aggressive behaviour if unfamiliar people or animals appear on their territory.*

expensive than some other forms of poultry, but, with specialist breed clubs and regional poultry clubs as contact points, reliable breeding stock can usually be found. The increasing

▼ *Geese love to preen, and in doing so add tranquillity to the poultry yard.*

numbers of European breeds now being kept reflects regional interest in local breeds. Some of these are small and compact breeds that are useful for owners with large gardens or small paddocks. Many will want to keep geese, if not as pets, then as loyal companions that interact with their owners and friends more readily than any other domestic fowl.

Exhibition forms of the larger, slow-maturing birds may still have a place in future breeding programs.

CARING FOR GEESE

Geese are long-lived birds, with some breeds thriving for 20 years or more, so investing in an ornamental bird for your garden is a long-term commitment. Geese bought and reared for their meat have a short lifespan, since it is not economically viable to keep them beyond 23 weeks.

Geese can be aggressive, and may adopt an intimidating attitude to strangers. They are also protective of their brooding spouse, goslings and nest, though they will generally settle down to become companionable, if noisy, garden neighbours. Pure-breed geese have a variety of characteristics. Lightweight Chinese geese are alert, with loud voices that make them good watchdogs, while the far heavier and more statuesque Toulouse will provide a more placid companion.

Buying geese

When it comes to purchasing stock, people wishing to keeping pure breeds should choose stock from an established breeder. Breeding groups, comprising of a gander and two to four geese, will need time to get used to each other and to their surroundings. This means that the best time to purchase them is in autumn or early winter so they have time together before spring.

▲ *Some geese have been known to challenge a fox; nonetheless, all geese will need to be securely housed at night. They require a big door as they will sometimes have to be herded in at dusk. Geese love a clean bed of straw that they will soon get messy, so to assist in cleaning a low house, a removable roof can sometimes be fitted.*

▼ *A plentiful supply of fresh, clean drinking water must be provided for geese, along with enough to spare for them to bathe their heads and necks at all times.*

▲ *Geese are good grazers, and when grass is both plentiful and nutritious they will live and fatten off good grass. During very dry weather and the winter months they require extra cereals, changing to a proprietary ration complete with minerals and other nutrients during and just before the breeding season.*

Housing geese

Nearly all geese are vulnerable to attack by predators at night and should be housed in a secure pen or similar fox-proof run. These adaptable birds can be kept in any type of shed or building as long as it is secure. Rather than perching, geese spend the night at floor level, so a wide door-way will be useful when persuading them to retire. Smaller units are easiest to clean if they have human access to remove soiled bedding.

Geese love water, but it is not necessary to provide a pond for swimming. They will enjoy a small tub, which should be adequate for bathing; geese do not have to spend as much time on the water as many of us think. Make sure fresh water for drinking is provided every day.

Feeding geese

Geese eat a totally vegetarian diet that for much of the year consists of grass. Good quality, well-managed grass and clover can provide a high proportion of the total nutritional needs of adult and growing geese. However, remember that geese eat huge quantities of grass, and much of it after the main grass-growing season is over, leaving large areas of soil exposed. The resulting exodus from their bodies also means that paths may be covered in pea-green goo.

A regular feed of proprietary goose feed will help to ensure healthy goslings that hatched early will be able to take advantage of any later flush of spring grass.

Baby and young goslings, particularly those being intensively brooded, need four or five feeds a day of damp, crumbly mash. This should be a specialist waterfowl formula rather than chick feed, which may contain unsuitable additives.

Fattening the goose

Goslings hatched in spring were traditionally fattened on grass and finished on gleanings from fields in time for Michaelmas (29 September). Such birds were known as "green

geese" and were prized for their tender flesh. Nowadays most of the larger flocks of geese used to supply the specialist Christmas market will have been supplied as goslings produced from specialist breeders. Such goslings are generally from breeders using advanced strains of

▼ *Geese that are to be sold for their meat are fattened on grain ready for market.*

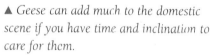

▲ *Geese can add much to the domestic scene if you have time and inclination to care for them.*

good laying geese mated to heavy, quick-growing ganders. These young birds are started on specialist gosling mash or crumbs, then feed on the succulent, high-protein grass of spring and early summer. The later drier, higher-fibre grass may make up much of the maintenance diet of adult geese, but growing goslings and young fattening geese will quickly require this to be supplemented by a grain-based mix.

Fattening geese require unstressful conditions. They may be left on grass in a restricted space, so that they conserve their energy. Some owners use electric fencing to pen their geese. If open ground becomes waterlogged the geese may be moved to a yard, and transferred to a house at night. These sociable birds may go off their feed if some of their number are removed early. Avoid changes to their regime that may unsettle them.

KEEPING TURKEYS

No other member of the poultry family interacts with its owners quite like a turkey. These are large, friendly and docile creatures that add interest to a larger garden and can add variety to a domestic poultry fowl collection, in much the same way as they did to traditional farmyards.

Turkeys may be a much rarer sight and sound in the neighbourhood than garden hens and roosters, but people keep them for much the same reasons. Turkeys are large birds, with several species available. As a breed, turkeys can make unusual pets, and like chickens, there are rare varieties, each with their own distinctive characteristics from which to choose. Like chicken fanciers, there are turkey fanciers, who select and breed turkeys for conservation and exhibition purposes. However, being larger and with a greater risk of breaking feathers, not to mention being unwieldy to transport, turkeys are much less prominent at poultry exhibitions than other types of fowl.

▼ The American Bronze turkey is farmed for its meat. The various forms of this breed have added size and their distinctive character to farmyards for hundreds of years.

For some people, the opportunity to rear the bird that they will eat at Christmas or Thanksgiving may be appealing. Birds can be bought in as young, known as poults, at one day old, and may be reared in the garden in much the same way as chickens. Provided the source of origin is reputable, you will have the security of knowing how the bird has been reared and on what it has fed, helping to allay concerns about animal welfare and any contamination of the food chain. However, turkeys are classed as livestock by some local authorities and permission may be required in order to keep them.

Male turkeys, known as stags or toms, make a wide range of vocal

▼ Turkey eggs can be used for cooking and baking in much the same way as a hen's eggs, though they are never produced on the scale of hybrid hens reared specifically for this purpose.

noises, particularly gobbling, so if you live in close proximity to neighbours, it may be sensible to discuss your ideas with them before going ahead and making a purchase. Keeping your neighbours' goodwill is always preferable. Since all fowl require care and attention for 365 days of the year, asking your neighbours to help provide emergency cover, should you need it, will be easier if they have no objections to the fowl in the first place. In many countries, turkeys have to be slaughtered by a qualified butcher, so if you are keeping a bird in order to eat it, ensure that you know how far you may have to travel with it in order to have it prepared for the oven. Turkeys can also be large and expensive pets to feed.

Turkey eggs

Turkeys have never been bred for egg production in the way that hens have, but some strains, notably the Buff turkey breed, are good egg-layers. Turkey eggs are rarely seen for sale, since relatively few are produced when compared to the numbers of hen and duck eggs laid.

Turkey meat

The original wild, native American, bronze-coloured turkeys would have had boat-shaped bodies, and be well flavoured, but by modern standards, the meat would be regarded as tough. Today, turkey is farmed commercially to supply meat for human consumption as a cheap source of protein. Since turkeys are closely associated with the commercial meat trade, their shape and size have been directly affected by market demand.

▲ *Good management and an attention to detail helps ensure that turkeys being reared in groups on straw remain both healthy and contented.*

In order to produce birds that will develop a large quantity of meat for very little feed, and which sell at an economic price, breeders have consistently, over time, selectively bred from fowl that display the most desirable characteristics.

Several strains or varieties of Bronze or Black turkeys evolved between the 1850s and 1950s to meet market demand. The American Bronze has Mexican ancestry; the wild bird was taken to Spain when Mexico became Spain's territory. The breed was taken back to America by migrants from Europe centuries later. These European-bred turkeys were crossed with American wild turkeys to produce larger birds that became known as the Bronze. The new breed was first standardized as the Cambridge or Standard Bronze. Early crossing of the new breed with giant American birds produced a new type known as the American Mammoth or Broad-Breasted Bronze, which

weighed up to 18.4kg/40lbs. This larger type could not mate naturally, and only survived via artificial insemination in order to produce fertile eggs. These breeds were developed to such an exaggerated size that most were only suitable for intensive indoor production. However, the arrival and development of these fast-growing strains enabled the industry to implement intensive or factory farming methods, with the

product often sold frozen and far more cheaply than any other similarly produced broiler poultry meat.

Trade and consumer preference for birds with a large carcass and white breast meat led to the demise of utility flocks of the old pure breeds, which could not compete on the same scale as intensively produced meat. By the 1980s, consumer preferences began to change, and over the following decades the discerning diner began to demand turkey meat with more flavour, while animal welfare campaigners demanded that birds be reared in more humane circumstances.

Small-scale pure-breed farming

Norfolk Black and American Bronze pure-breed turkeys that remained in the hands of traditional, small-scale farmers and breeders throughout the decades of mass production were to

▼ *Traditional free-range Bronze turkeys being fattened will find some of their own feed. Spending time investigating every part of the run in the pursuit of food can help prevent boredom, which can lead to feather pecking and aggression.*

become the basis of the growth in farming traditional breeds. These breeds, having taken longer to mature to a less exaggerated size, are considered better quality, with superior meat flavour. In addition, they are reared in a more humane way than those under the intense factory farming system.

These breeds are also being developed to meet the same market requirements, competing with intensively farmed products by appealing to consumer conscience. A growing demand for quality Christmas turkeys may see a return to medium-sized flocks being reared on less specialized farms.

Small-scale turkey farmers vary considerably in the scope of their interest, and each will have a management regime that reflects their commercial requirements.

Since turkeys lay eggs less reliably and in much smaller quantities than hens, in order to ensure a regular supply of birds to grow on farmers may buy in poults at one day old,

▼ *A free-range turkey farm provides good conditions for its stock.*

▲ *Turkey-keeping will require a higher standard of stockmanship and management skills than nearly any other form of poultry keeping.*

supplementing rather than relying on breeding their own turkeys. Poults are best purchased as two mixed batches of different types, each timed to finish at different weights, sourcing them from a specialist breeder who will advise on optimum age and weight as well as growth rate of available stock.

Turkey franchises

Many medium-sized producers run franchises. Working within a franchise arrangement may not suit every producer, since the end product still sells to a smaller, though growing, niche market. Producing a top-quality turkey meat requires the most selective breeding, feeding and free-range production. There must be close co-operation with the breeders who supply the franchise to ensure the poults arrive at the right time to meet the required weights.

A franchise will ensure that the smallest hens of the batch reach 1.7kg/3½lbs (oven-ready weight) at 10 weeks and that the biggest stags finish at 11.8kg/26lbs (oven-ready weight) at 24 weeks. The turkeys are killed, processed and dry-plucked a short distance from the rearing farm.

Intensive turkey farming

Much large-scale turkey production is based on a factory system that sees birds intensively housed throughout their lives. As they reach optimum weight, they are killed and kept frozen until they are marketed.

Most of the available day-old, usually white poults that form the basis of the industry will have been selectively bred for a monocropping system. The larger hatcheries will often keep flocks of laying females that are artificially inseminated from heavier males. As it is not viable to maintain this type of breeding setup to produce poults all year round, most small-scale producers will find themselves competing to purchase poults during the same summer period. Poults can be reared under broody hens, but most producers will opt for the heat lamp system used for day-old and growing chicks. While turkey poults require marginally higher brooding temperatures during their early life, as they mature to become fully feathered, many will require less shelter during their remaining lifespan.

Conservation of old turkey breeds

Within this specialized market, smaller producers and exhibition breeders have adopted differing breeding and management practices to those of industrial producers.

The purest strains of the old breeds of turkeys are selected to finish growing at a specific weight, with some slow-maturing Bronze males increasing to a huge size at eight months old. The females mature at a younger age and to lower weights. At the other end of the size scale are lighter breeds, such as the Beltsville White, that mature as plump tiny birds at less than 16 weeks old. Examples of the old pure breeds are still occasionally seen at poultry shows. The fact that these breeds have survived means that they may still have a role to play in the future of commercial poultry production.

Enthusiastic conservationists often farm the large Bronze and White varieties as well as breeds such as the Norfolk Black, steering a path between over-developed dimple-breasted types while still providing a rounded, full-flavoured product. Most of these producers rely on bought-in chicks or poults.

Buying and caring for turkeys

People who favour rearing pure-breed turkeys will purchase poults or eggs from a specialist breeder. If you intend keeping turkeys they will need an adequate space in which to roam.

◄ *Turkeys will relish left-overs, overripe fruit, and fruit and vegetable peelings.*

A breeding pair will require 8sq m/ 90sq ft of space as a minimum. A 2m/6.5ft-high fence should be sufficient to prevent the birds from escaping. It is not advisable to allow turkeys to roam free in the garden as they will destroy plant growth and blooms. Turkeys are naturally curious birds, so make sure that the penned area is free from all potential hazards that could harm them. Turkeys kept outside will find their own grit to aid digestion. Be sure to provide those kept indoors with sand or fine gravel.

Feeding turkeys

Grass makes up a small proportion of a free-range turkey's diet. The majority of the feed regime consists of grain, vegetable protein, pulses and dried grass. This diet will provide essential minerals and trace elements without recourse to drugs, genetically modified feed or animal protein. This ensures that the meat is well flavoured.

▼ *Narragansett turkey.*

▼ *Beltsville white turkey.*

▼ *Norfolk Black turkey.*

INDEX

▼ *A Fayoumi hen.*

▼ *A Croad Langshan.*

▼ *A Blue Andalusian.*

▲ *A Jubilee Indian Game fowl.*

▲ *A Rumpless bantam.*

ACKNOWLEDGEMENTS

The publishers and author would
like to thank the following for
allowing photography:

The National Federation of
Poultry Clubs show at Stafford.
The help of members and judges
at the Poultry Club of Great
Britain. Affiliated shows run by
Arun Valley Poultry Fanciers
Society; Hants and Berks
Poultry Fanciers Society;
Kent Poultry Fanciers Society;
Norfolk Poultry Club; Reading
and District Bantam Society; and
Surrey Poultry Society.

▼ *A Sicilian Buttercup.*